Agrometeorology

Edited by Ram Swaroop Meena

Published in London, United Kingdom

IntechOpen

Supporting open minds since 2005

Agrometeorology
http://dx.doi.org/10.5772/intechopen.87446
Edited by Ram Swaroop Meena

Contributors
Nukshab Zeeshan, Nabila ., Ghulam Murtaza, Zia Ur Rahman Farooqi, Khurram Naveed, Muhammad Usman Farid, Ram Prakash Yadav, Suresh Chandra Panday, Jitendra Kumar, Jaideep Kumar Bisht, Vijay Singh Meena, Mahipal Choudhary, Shyam Nath, Manoj Parihar, Rajendra Prasad Meena, Valentin Stoyanov Kazandjiev, Kingsley Eghonghon Ukhurebor, Godfrey Shem Juma, Festus Kelonye, Ponsian T. Sewando, Marcos Paulo Gomes Gonçalves, Mohammed Rahman, Akinori Yamamoto, Jane Asiyo Okalebo, Taity Changa, Shaokun Wang, Boubakeur Guesmi, Felix Gemlack Gemlack Ngasoh, Constantine C. Mbajiorgu, Matthew B. Kamai, Suad Al-Manji, Gordon Mitchell, Amna Alruheili, Salisu Lawal Halliru, Abubakar Abdullahi Bich, Aliyu Shu'aibu Muhammad, Veska Anastassova Georgieva, Gideon Onyekachi Okoro

First published in London, United Kingdom, 2021 by IntechOpen
IntechOpen is the global imprint of INTECHOPEN LIMITED, registered in England and Wales, registration number: 11086078, 5 Princes Gate Court, London, SW7 2QJ, United Kingdom
Printed in Croatia

British Library Cataloguing-in-Publication Data
A catalogue record for this book is available from the British Library

Additional hard and PDF copies can be obtained from orders@intechopen.com

Agrometeorology
Edited by Ram Swaroop Meena
p. cm.
Print ISBN 978-1-83881-174-7
Online ISBN 978-1-83881-175-4
eBook (PDF) ISBN 978-1-83881-176-1

For EU product safety concerns:
IN TECH d.o.o., Prolaz Marije Krucifikse Kozulić 3, 51000 Rijeka, Croatia, info@intechopen.com or visit our website at intechopen.com.

Meet the editor

Dr. Ram Swaroop Meena is an assistant professor in the Department of Agronomy, Institute of Agricultural Sciences, Banaras Hindu University (BHU), Varanasi, Uttar Pradesh. He was awarded the Raman Research Fellowship by the Ministry of Human Resource Development (MHRD), Government of India (GOI). Dr. Meena completed his postdoctoral research on soil carbon sequestration under Padma Shari Prof. Rattan Lal, World Food Prize 2020 Laureate, Director, Carbon Management and Sequestration Center (CMASC), Columbus, Ohio, USA. He has supervised twenty-five postgraduate and six doctoral students and has eleven years of research and teaching experience. He is currently working on the three externally funded projects (DST, MHRD, ICAR) with one patent. Dr. Meena has published more than 110 research and review papers, 65 book chapters, and 20 books. He has worked as an expert for the National Council of Educational Research and Training (NCERT), MHRD, and GOI. Dr. Meena has also contributed to several agricultural extension activities, trainings, meetings, workshops, and more.

Contents

Preface

Agrometeorology involves climatology, meteorology, agrology, biology, and hydrology. It requires a multidisciplinary range of data for operational applications and research. Current concerns for the sustainability of agroecosystems in different sectors of agriculture have heightened awareness for the careful use of natural resources. For proper and efficient use of soils and plant/animal genetic material, knowledge of the role of climate is essential. A more inclusive definition of agricultural meteorology deals with water, heat, air, and related biomass development, above and below ground, in the agricultural production environment, including the impact of pests and diseases that also depend on these factors. Using remote sensing (RS), Global Positioning Systems (GPS), and Geographic Information Systems (GIS) to develop crop inventories helps decision-makers and planners with best practices for agricultural management. The data analysis of crop details using various geospatial technologies fills in gaps in statistical agriculture research and provides reliable, replicable, and efficient methods for generating agricultural statistics.

This book includes a series of pictures that illustrate the agrometeorological features fundamental to the understanding of the subject. It contains thirteen chapters organized in order of increasing complexity. The chapters discuss the definitions, aims, scopes, and importance of agricultural meteorology and various meteorological parameters such as radiation, air temperature, air pressure, winds, humidity, and evaporation/evapotranspiration.

This volume is designed for students and researchers who desire to further devote their careers to the protection of natural ecosystems and increasing agricultural productivity and sustainability.

Ram Swaroop Meena
Department of Agronomy,
Institute of Agricultural Sciences (BHU),
Varanasi, UP, India

Chapter 1

Prediction of Crop Yields under a Changing Climate

Godfrey Shem Juma and Festus Kelonye Beru

Abstract

The impact of increasing climate variability on crop yield is now evident. Predicting the potential effects of climate change on crops prompts the use of statistical models to measure how the crop responds to climate variables. This chapter examines the usage of regression analysis in predicting crop yield under a changing climate. Data quality control is explained and application of descriptive statistics, correlation analysis and contingency tables discussed. Methodological aspects of crop yield modeling and prediction using climate variables are described. Estimation of yield via a multilinear regression approach is outlined and an overview of statistical model verification introduced. The study recommends the usage of regression models in estimating crop yield in consideration of many other externalities that can contribute to yield change.

Keywords: crop, yield, prediction, climate change, regression model

1. Introduction

In this chapter, we describe an experimental approach that can be employed in predicting crop yield in a changing climate. An introductory applied approach to linear statistical modeling and correlation analysis is examined.

Climate change is now evident with well documented socio-economic impacts that will affect food production [1, 2]. The decline in food production corresponding to reduction in crop yields can be investigated using statistical models [3, 4]. While climate related factors can affect yield of crop, there are other externalities that can impact on yield production that include the quality of soil, usage of commercial fertilizers or organic manures and residual effects of chemical substances in soils [5, 6].

Figure 1 below shows the projected changes in crop yield due to the impacts of increasing climate variability.

Increasing climate variability and associated uncertainties, its impacts on food production and general livelihoods prompts the usage of prediction models to estimate future food production for early warning and planning. Projection of crop yield in a changing climate has been identified with uncertainties that are continuously being reduced by improvement in climate parameter response functions, including temperature [7]. Developing countries have also been identified with weaker monitoring and reporting of crop health which can lead to absence of early warning systems and slow responses to droughts and potential food shortages [8]. The prediction models employed are broadly classified into statistical or dynamic (mechanistic). However, the modeling has been in some instances enhanced by artificial neural network technology that has been applied to generate regional

IntechOpen

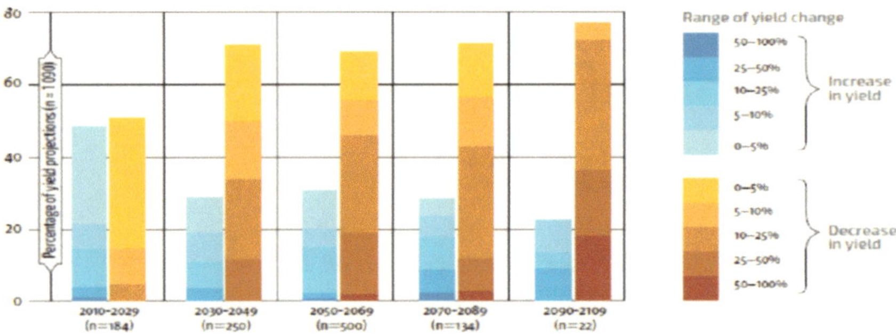

Figure 1.
Projected changes in crop yield associated with increasing climate variability (source: [7]).

time series of crop yield utilizing highly resolved output of the global and regional models. The Global Circulation Models (GCMs) give coarse output of important climate parameters preferably applicable on larger spatial scales while the Regional Circulation Models (GCMS) give fine scale resolutions of GCMs on shorter spatial scales. Where the variables of interests cannot be described by standard linear models, nested-error regression model having both fixed and random effects are usually applied to include Monte Carlo simulation methodology in order to enhance representative precision. Nested-error modeling is better at regional spatial scales and performs poorly at large scale spatial coverage [9].

Application of computing technology has seen historical climate data sets of at least 30 years of a given crop used with artificial neural network technology to investigate, simulate and predict historical time series of crop yields in climate zones over regions. Resultant neural networks are trained with data sets of selected climate zones and tested against an independent zone in order to enhance the power of crop yield predictability [10]. A combination of neural networks and fuzzy set theory has also been applied to construct Fuzzy Neural Network (FNN) and Granular Neural Network (GNN) that have been used for predicting crop yields over different spatial locations with inputs from spot vegetation cover data [11].

Assessment of vegetative cover using standardized vegetative indices has also been adopted to give estimates of crop yield over regions either separately or in combination of the above approaches. In this method, easily measurable proxies are applied that include Normalized-Difference Vegetation Indices (NDVI), Green vegetation indices (GVI), Soil Adjusted Vegetation Indices (SAVI), Back-propagation Neural Network (BPNN) that are positively correlated with crop yield [12]. Statistical models that combine the vegetative and thermal indices from satellite data have performed better in predicting crop yield compared to those that are based on vegetative cover indices alone [13]. While mechanistic models have been applied alongside statistical models, the later have been able to reproduce key features of crop responses to warming and precipitation changes using a process-based model approach [3, 14]. The Crop Environment Resource Synthesis (CERES) model has been applied as a decision making tool in crop yield estimation [15] while PRECIS (Predicting REgional Climate for Impact Studies) has been used to assess the impacts of climate change on crop yield [14].

Therefore, the need to continuously enhance understanding on yield estimation and prediction in a changing climate is continuous and cannot be over emphasized. It is in this back drop that this study makes attempts to adopt a simplified linear statistical approach applicable in crop yield estimation. Basic statistical description is defined and methodology discussed. Hybrid models that are both statistical and mechanistic, integrated by neural network technology based on multiple variables of climate and crop physiological importance are found to have higher crop yield predictive power.

2. Statistical determination of crop yield

2.1 Descriptive statistical analysis

This entails describing data in statistical summaries meaningfully without making conclusions beyond the data. Measures of central tendency and measures of spread are widely adopted in describing data. The former involves the determination of the mean, median, mode, skewness and kurtosis while the later includes variance and standard deviation. These parameters can be used to describe climate and crop yield data as a preliminary approach alongside data quality control.

Data quality control involves approaches that are used to detect defections and inconsistencies in data sets. Various methods are used to determine the quality of climate data including linear regression approaches. In one such method, a single mass curve technique is applied where cumulative values of climate variable are plotted against a linear scale. The tendency of the resulting curve to shift towards linearity is identified with better quality data. This method is also called the data homogeneity test. Data that fails this test or data with more than 10 percent of missing values is judged to be of poor quality and not fit for inferential statistics.

2.2 Crop yield modeling and prediction

Statistical models have been applied in predicting crop yield and their ability to accurately predict yield responses to changes in mean temperature and precipitation has been determined by process-based crop models. Prototype models include Crop Environment Resource Synthesis (CERES) that can be applied to a crop to simulate corresponding yield and can be used for projecting future yield responses, with their usefulness higher at broader spatial scales [16]. *Mechanistic* models are also used alongside statistical models to predict crop yields [17]. Crop Yield Simulation and Land Assessment Models (CYSLAM) are applied to model the interaction of environmental variables, physiological responses, inputs, yields and land management mechanistic simulations of crop yield [17].

Yields constrained by radiation and temperature within 10 day periods (dekads) are initially estimated in order to account for effective rainfall, evapotranspiration, percolation, and soil moisture. The procedure is followed by a simulation of crop/soil water balance through the cycle of crop growth accounting for periods of moisture stress and consequently, estimation of crop yield [17]. The moisture-dependent yield is adjusted for nutrient supply, toxicities and drainage conditions of the soil [17]. However, validation of modules for moisture limited yield, nutrient yield and radiation and temperature limited yield is carried out separately in comparison with historical crop data.

2.2.1 Multilinear regression yield estimation

Single mass curve technique is used for data quality control where Cumulative values of data are plotted against a temporal scale. The nature and variability of climate elements is determined including the mean, skewness, standard deviation, students' t-test and correlation analysis. Trend is determined by dividing the data into two sets of equal length, and the difference in the means of the two sets is tested using the t-test [4].

The Relationship between crop Yield and Variations in Climatic Elements is carried out by Correlation Analysis, the degree of relationship between at least two variables. The Pearson's correlation coefficient (r) is used to determine the correlation between the climate elements and the crop yield according to the following expression (1) below:

$$r - \frac{\sum_{i-1}^{N}(x_i - \bar{x})(y_i - \bar{y})}{\left[\sum_{i-1}^{N}(x_i - \bar{x})^2 \sum_{i-1}^{N}(y_i - \bar{y})^2\right]^{1/2}} \tag{1}$$

where

N => Total number of observations

$\bar{x}$ => Mean of the variable 'x'

$\bar{y}$ => Mean of the variable 'y'

Hypothetical approach is employed to test for statistical significance of the degree of association in (1) above. The null hypothesis that the correlation is zero and the alternative hypothesis that the correlation is nonzero is assumed. In this case, if the null hypothesis is valid, the relevant test variable (t) from Eq. (2) is a realization of student (t) random variable with mean (zero) and (n) degrees of freedom. P values are computed where $p < 0.05$ prompts the probability of rejecting the null hypothesis and vice versa. The student t-statistic can be used as given by the Eq. (2) below:

$$t - r\sqrt{\frac{n-2}{1-r^2}} \tag{2}$$

Multiple Linear Regression Analysis gives models that involve more than one independent variable and one dependent variable. This gives an analytical model, which is used to develop a model for predicting crop yield from climatic elements at various time lags.

This relationship is given by the Eq. (3) below:

$$Y = ßo + ß_1 X_1 + ß_2 X_2 + \cdots + ß\kappa X\kappa + \varepsilon \tag{3}$$

Where βs are coefficients, Xi are the predictors, Y is the crop yield (predictand) and β0 is a constant.

Climate variables are the independent variable while yield is the dependent variable. The data is imported into statistical analysis software (SYSTAT or R) where regression analysis is carried out in order to get coefficients $ßo$, $ß_1$ and $ßn$. These values are fitted in a multiple regression model of the form (3) above.

The regression model for predicting crop yield is arrived at via a series of enhancing steps where the initial step entails all climate variables specified in the data file. Climate variable with a p-value greater than 0.05 are judged as statistically significant in the model at 95% confidence level. The second step is a repeat of the first one excluding the non-significant variable in the data file. A model that entails the statistically significant climate variables is specified and adopted.

The third step entails plotting the model residuals with keen interest on the normal Q-Q plot to detect the outliers. Where outliers are detected, the model has to be "re-built" without outliers. The error terms which are the differences between the observed value of the dependent variable and the predicted value are called residuals. The final outlier-less model is specified with the following key assumptions namely: Homoscedasticity of residual based on equal variance; Normality of residual; Leverage based on distance of plots to the center and the cook's distance; Positive variance and non-perfect multicollinearity. Homoscedasticity is defined by a scatter plot and assumes equal distribution of the residuals. Normality of residual assumes that the regression follows a normal distribution. Cook's distance provides an idea on influential data points that are worth checking for validity. Non-perfect multicollinearity occurs when one of the regressors is highly correlated with, but not equal to a linear combination of other regressors.

Contingency table can be used for verifying the model. Data is split into two data sets where one set is used in training the model in a statistical analysis tool (e.g. SYSTAT). Model verification statistics including percent correct (PC), Post Agreement (PA), False-Alarm Ratio (FAR), critical success index (CSI), probability of detection (POD), bias, and Heidke's skill score (HSS) were determined.

The methodology was applied by [4] in their assessment of crop yield over Nandi East Sub-County in Kenya.

3. Conclusion and recommendation

This book chapter described a basic regression approach applied in predicting crop yield in a changing climate. Introductory concepts of descriptive statistics, data quality control, correlation analysis and multilinear modeling are discussed. A typical regression method to estimate yield is examined. The study concludes that crop yield prediction and estimation is married with uncertainties of both natural and anthropogenic nature and requires continuous improvement with more focus on externalities that affect crop yield. This study recommends hybrid models that are both statistical and mechanistic, integrated by neural network technology based on multiple variables of climate and crop physiological importance.

Acknowledgements

We acknowledge the contribution of the late professor Joseph Mwalichi Ininda of the Department of Meteorology, University of Nairobi for technical support.

Thanks

Much thanks to the office of the deputy vice chancellor, PPRI; Kibabii University for much support.

Author details

Godfrey Shem Juma[1*] and Festus Kelonye Beru[2]

1 Department of Mathematics, Kibabii University, Kenya

2 Department of Biological and Environmental Sciences, Kibabii University, Kenya

*Address all correspondence to: godfrey.juma@kibu.ac.ke

References

[1] Stocker TF, Qin D, Plattner GK, Tignor M, Allen SK, Boschung J, et al. Climate change 2013: The physical science basis. Contribution of working group I to the fifth assessment report of the intergovernmental panel on climate change. 2013;**1535**

[2] Kumar M. Impact of climate change on crop yield and role of model for achieving food security. Environmental Monitoring and Assessment. 2016;**188**(8):465

[3] Lobell DB, Burke MB. On the use of statistical models to predict crop yield responses to climate change. Agricultural and Forest Meteorology. 2010;**150**(11):1443-1452

[4] Sitienei BJ, Juma SG, Opere E. On the use of regression models to predict tea crop yield responses to climate change: A case of Nandi east, sub-county of Nandi county, Kenya. Climate. 2017;**5**(3):54

[5] Uzen, N., Cetin, O., & Unlu, M. (2016). Effects of domestic wastewater treated by anaerobic stabilization on soil pollution, plant nutrition, and cotton crop yield.

[6] Asare, E., & Scarisbrick, D. H. (1995). Rate of nitrogen and sulphur fertilizers on yield, yield components and seed quality of oilseed rape (*Brassica napus* L.). Field Crops Research, 44(1), 41-46.Environmental monitoring and assessment, 188(12), 664.

[7] Dwivedi, A., Naresh, R. K., Kumar, R., Kumar, P., & Kumar, R. (2017). Climate smart agriculture. no. December.

[8] Petersen LK. Real-time prediction of crop yields from MODIS relative vegetation health: A continent-wide analysis of Africa. Remote Sensing. 2018;**10**(11):1726

[9] Herrador M, Esteban MD, Hobza T, Morales D. A modified nested-error regression model for small area estimation. Statistics. 2013;*47*(2):258-273

[10] Heinzow, T., & Tol, R. S. (2003). Prediction of crop yields across four climate zones in Germany: an artificial neural network approach (No. FNU-34).

[11] Savin IY, Stathakis D, Negre T, Isaev VA. Prediction of crop yields with the use of neural networks. Russian Agricultural Sciences. 2007;*33*(6):361-363

[12] Panda SS, Ames DP, Panigrahi S. Application of vegetation indices for agricultural crop yield prediction using neural network techniques. Remote Sensing. 2010;*2*(3):673-696

[13] Holzman ME, Rivas R, Piccolo MC. Estimating soil moisture and the relationship with crop yield using surface temperature and vegetation index. International Journal of Applied Earth Observation and Geoinformation. 2014;*28*:181-192

[14] Erda L, Wei X, Hui J, Yinlong X, Yue L, Liping B, et al. Climate change impacts on crop yield and quality with CO2 fertilization in China. Philosophical Transactions of the Royal Society B: Biological Sciences. 2005;**360**(1463):2149-2154

[15] Popova Z, Kercheva M. CERES model application for increasing preparedness to climate variability in agricultural planning–calibration and validation test. Physics and Chemistry of the Earth, Parts A/B/C. 2005;**30**(1-3):125-133

[16] Estes LD, Bradley BA, Beukes H, Hole DG, Lau M, Oppenheimer MG, et al. Comparing mechanistic and empirical model projections of

crop suitability and productivity: Implications for ecological forecasting. Global Ecology and Biogeography. 2013;**22**(8):1007-1018

[17] De Wit PV, Tersteeg JL, Radcliffe DJ. Crop yield simulation and land assessment model for Botswana (CYSLAMB). Part I theory and validation. FAO/government of Botswana. Land resource assessment for agricultural land use planning project TCP/BOT/0053. Field Document. 1993;**2**(72):1993

Chapter 2

Climatic Variation and Its Impacts on Yield and Water Requirement of Crops in Indian Central Himalaya

Ram Prakash Yadav, Suresh Chandra Panday, Jitendra Kumar, Jaideep Kumar Bisht, Vijay Singh Meena, Mahipal Choudhary, Shyam Nath, Manoj Parihar and Rajendra Prasad Meena

Abstract

Climate is most important factor affecting agriculture, and issues related to climate and its implications have attracted attention of policy makers globally. The farm sector, particularly marginal ecosystems in mountains are vulnerable because of unpredictable variation and severe sink limitations. Efforts to impart resilience to farm and its allied sector are an urgent need. The climatic parameters play very important role to determine type of crops, cattle rearing and the life style adopted by the people. Moreover, weather has a significant impact on crop growth and development. Weather plays a vital role and affects the production and productivity of the crops. According to an estimate, weather contributes 67% variation in productivity and rest of the factors (soil, nutrient and management practices etc.) accounts for 33%. Therefore, there is a need of in-depth analysis of each meteorological parameters and identification of their trend over the years in order to identify and adapt suitable agriculture practices, better adaptable crops, varieties and their duration, time of field preparation, sowing time and irrigation as per the climatic conditions of the region. This will lead farming community to plan strategies of agriculture operation to obtain optimum yield. The climatic data from the meteorological observatory of ICAR-VPKAS, Hawalbagh located at mid hill condition (1250 m amsl) were analyzed for different periods (annual, seasonal, monthly, weekly). It was revealed that rainfall is decreasing over the years but significant ($P < 0.05$) decrease was recorded at mid hills. The maximum temperature is increasing significantly ($P < 0.05$) during post-monsoon and winter season however decreasing in monsoon season whereas minimum temperature is decreasing round the year. These changes in rainfall and temperatures are affecting production and productivity of the crops, as hills are largely rainfed. In terms of crop water demand, there is no need to apply irrigation during the rainy season except the transplanted rice. However, during the winter season as there is more than 60% of water deficit to irrigate the crops. The proper understanding of climate is necessary to bring sustainability in hill agriculture by adjusting crop sowing window and other operations as per suitability of the climate.

IntechOpen

Keywords: agroforestry, climate change, Himalaya, meteorological data, productivity, sustainability

1. Introduction

Climate determines nature and productivity of farming systems' and allied farm enterprises. The climate over the Himalayan ranges varies from place to place due to complexities of the local relief features and type of weather systems. Weather is defined as the state of the atmosphere at a specific time and place. It defines the physical conditions of the atmosphere with respect to wind, temperature cloudiness, moisture, rainfall pressure and other parameters such as sunshine hour, evaporation etc. Climate pertaining to a region or a place can be defined as the sum of the weather conditions prevailing over a place over the years. In other words, climate represents average weather conditions over a long period. It comprises not only those conditions that can be obviously described as near average or normal, but also the extremes and all the variations. According to an estimate, weather contributes 67% of the variations in productivity and rest by the other factors viz., soil nutrient and management practices [1]. The climate affect the crop production, jeopardize the livelihood of people, and induced animal starvation. The recent severe drought of 2002 affected 300,000,000 people and monetary losses US $910,721,000 [2]. It is well known fact that there is little control over the climate and there is a very little chance to manipulate the weather. Therefore, there is a need of identification and adaption of suitable crops and varieties, as per the climatic conditions of the region.

The potential of climate as an agricultural resource has not been fully utilized and realized. As a result, several crops are grown traditionally without any knowledge of their suitability. Thus, on one hand, poor production and on the other, much of the production potential of these vast resources go untapped. However, it is inevitable to make adjustment with weather to harness the maximum benefit from available climatic resources. Therefore, knowledge of agro climatology of a region is a valuable tool for crop planning. The knowledge of climatic conditions also helps in identification and selection of forest trees, species and grasses. The proper knowledge of climate and its relationship with severity of insects and pests, diseases will be helpful for managing them and minimizing the losses. Therefore, knowledge of the climate is necessary to utilize resources efficiently for maintaining the production and regeneration capacity. As we know that, there is little control over the climate and there is very little chance to manipulate the weather. Therefore, there is a need of in-depth analysis of each meteorological parameters and identification of their trend over the years in order to identify and adapt suitable agriculture practices, better adaptable crops, varieties and their duration, time of field preparation, sowing time and irrigation as per the climatic conditions of the region.

One of the major challenges of the 21st century is to ensure food security for the burgeoning population. It is more pertinent to India, where population growth rate is high and it may surpass the China's population in near future. Although consequent to green revolution India in general has made remarkable progress in production of food and fiber since 1970's. However, India has to progressively match the production with population growth rate still to ensure food security, it has to go for climate smart sustainable agriculture. Agriculture is even more important as a primary source of livelihood for majority of the world's workforce. However, livelihood systems that are based on agriculture may face growing risk of crop failure, frequent incidence of pest and diseases and loss of livestock due to climate change

[3]. Crop production is a complex phenomenon and depends on abiotic (soil, weather, plant etc.) and biotic factors (insect, pest diseases etc.). Sudden change and frequent variation of these factors during different crop growth stages can bring drastic change in crop production and productivity. Warming trends were reported in Indian climate [4]. The decline in production in rice and wheat due to climate variation was reported [5, 6] though received rainfall higher than the mean rainfall of rice and wheat. This chapter deals with the impact of climatic variation and changes on natural resources and agricultural productivity.

1.1 Agroforestry and climate change

In the welfare of our nation and its economy, value of forests and trees is foremost. The indirect benefits are more than direct benefits to us through the trees [7, 8]. It is true that if the tree is present on the earth, the water will be available; if water is there, food will be available and due to food, we will exist. Our resources were first developed in the shadows of the trees because the trees give us food, fuel, fodder, shade and timber and improve fertility of land as well as increase water availability, prevent soil erosion and help maintain environmental balance [8, 9]. At present, due to burgeoning population and dwindling resources, importance of agroforestry has increased [10]. Due to the development of new systems, in modern agriculture such as agro-forestry is need of the hour in place of traditional agricultural landscape without reducing food production [11, 12].

To meet the needs of growing population conflict between forest and agricultural land has also increased; industrialization has generated pressure on land for many other demands [13]. Therefore, it is necessary to produce both agricultural and forestry products on the same piece of land to meet out the increasing demand of different products [14]. For this, plantations should be done on the agricultural land. Agroforestry helps in removal of carbon and other greenhouse gases from the atmosphere to mitigate climate change, reducing the impact of climate change and reduces the vulnerability [8, 11, 12, 15]. The different options to mitigate climate change include increasing carbon reducing activities such as carbon sequestration, to reduce emissions of bio-energy and biofuels in productive activities. Agroforestry is not only a large and low cost opportunity to mitigate climate change but also provides other services to the community [16]. Tree collects carbon dioxide (CO_2) from the atmosphere and converts it into carbon molecules through photosynthesis in leaves, which is responsible for growth of tree. One kilogram of dry wood contains ~0.5 kg carbon. According to IPCC [17] tropical wet forest have carbon in above ground biomass of 65–430 t ha^{-1} and 44–130 t ha^{-1} in soil.

Tree provides leaves as manure for agriculture, help in farming directly and indirectly and provide food and fodder for human and animal [7, 8]. This may benefit from the same piece of land that help in self-catering as well as can reduce the risk, also helps in providing appropriate environment for crops and the market for products. Although this technique is very simple, still it provides complementary employment and a very small technical and economic cooperation is required. The tree is a blessing because it is very useful to the poor and provides various benefits and windfall profits due to different timings of reaping at low cost variation. The greatest significance of agroforestry is that the farmer's family can adopt it as a good activity without changing his business and trees are very cooperative in the agro-forestry [18]. Trees convert poisonous gas CO_2 into life saving oxygen (O_2). This action purify the air as well as help in preventing elevated temperature because trees absorb atmospheric carbon in the process of photosynthesis. In agroforestry, we should choose trees with short life cycle and fast growing in nature. India has diversified climate, different agricultural conditions and abundant wasteland in villages need to be considered while implementing various methods of forestry. Various agroforestry practices can be used based on climate of the place and utility, benefits and on need basis [7, 9, 19].

The provision of agro-forestry-based industries in National Agroforestry Policy (2014), will give a major boost to the agroforestry [20]. In 1981, our country's population was 236.7 million, which increased to 1.21 billion in 2011. However, there has been no increase in the country's geographical area. If population growth continues at this pace, by 2020 the population of our country will be more than 1.30 billion. Thus, the effects of increasing population and urbanization have impact on loss of forests, increased soil erosion and atmospheric pollution. Therefore, at present the agro-forestry tree planted portion will not only prevent soil erosion but also increase its productive power and will help to maintain environmental balance. Planting trees with crops will increase the returns from per unit land [19].

2. Climate and its variability at Mid hill Hawalbagh, Almora

The climate variability observed based on past years data of rainfall and temperature is summarized in next section. The rainfall variation showed that out of 55 years, the 24 years rainfall was below the normal rainfall. The annual rainfall showed decreasing trend and the rainy days with 50, 75 and 100 mm of rainfall are found increasing. Whereas, 25 mm rainfall rainy days are decreasing.

2.1 Rainfall characteristics

The precipitation is the primary input in the hydrological cycle and dominantly influences the complex hydrological phenomena. Precipitation is the main source of fresh water. Precipitation is any product of the condensation of atmospheric water vapor that falls under gravity. The main forms of precipitation include drizzle, rain, sleet, snow, graupel and hail. Rain is liquid precipitation, as opposed to non-liquid kinds of precipitation such as sleet, snow, and hail. The rainfall information is needed for management of natural resources and crop planning. Rainfall amount, intensity and its distribution are prime factors affecting plant growth, soil erosion and flood problems. This is more pertinent to state like Uttarakhand hills, which has only 10% area under irrigation. The distribution of rainfall in hills is very important because it not only influences hills but also downstream locations in plains. The excess rainfall may aggravate erosion in hills and floods in low-lying areas while deficit may cause drought and reduced river flow. Knowledge of rainfall pattern in a given geographical location enables the development of suitable strategies for agricultural planning and implementation.

2.1.1 Annual rainfall

Annual rainfall (1964–2018) data recorded at Agro-Metrological Observatory, Hawalbagh (Almora), showed variation from 650.8 mm in 1974 being lowest to 1496.0 mm in 1971 being highest with a mean of 994.8 mm. The standard deviation (SD) and coefficient of variation (CV) of annual rainfall were 208.7 mm and 21.0 per cent, respectively. It was observed that rainfall in 24, out of 47 years was higher than the mean. The annual average rainy days were 67 with SD 10.5 and CV 15.6 percent. The maximum number of rainy days (90) was recorded in the year 1977 with rainfall (1088 mm) and minimum (49) with rainfall (852.1 mm) in 1964. The highest annual rainfall (1496.0 mm) was recorded during 1971 in 85 days, while highest rainy days (90) received 1088 mm rainfall in 1977, respectively. Annual rainfall showed decreasing trend over the years.

2.1.2 Monthly rainfall

The mean monthly rainfall (1964 to 2018) is presented in **Table 1**. It is evident from table that each month received more than 130 mm rainfall during June to September. The highest rainfall (240.2-mm) with highest 14 rainy day was recorded in July followed by August with mean rainfall (210.0 mm) with average 13 rainy days. The monsoon months (June to September) regarded as effective months of rainfall or wettest months of the year. The average rainfall during October, November and December was 22.8, 6.2 and 20.3 mm respectively. The average rainfall during January, February and March months was 39.9, 50.7 and 42.4 mm with rainy days 3.0, 4.0 each for later two months 4.0 respectively. The rainfall of pre monsoon months *i.e.* the April and May was recorded 32.0 mm and 61.7 mm with rainy days 3.0 and 5.0 days respectively (**Table 1**).

2.1.3 Annual and seasonal rainfall and its variability

The annual rainfall showed decreasing trend (**Figure 1**). It was recorded that mean annual rainfall was 994.8 mm, highest annual rainfall (1496.0 mm) in the year 1971, lowest annual rainfall (650.8 mm) in the year 1974 and highest rainfall in single day (167.0) on 18th September 2010. Whereas, mean annual rainy days (68 days), highest rainy days in year (90 days) in the year 1977, lowest rainy days (49 days) in the year 1964 and 1966. The highest rainfall 72.3% occurred in monsoon season followed by summer (13.6%) winter (11.1%) and least in post monsoon (3%).

Month	Mean	lowest	Highest	Percent annual rainfall	CV (%)	Rainy days* (rainfall more than 2.4 mm)
January	39.9	0	113.3	4.0	79.7	3
February	50.7	0	153.4	5.1	76.0	4
March	42.4	0	195.4	4.3	89.1	4
April	32.0	0	153.2	3.2	88.7	3
May	61.7	0	179.2	6.2	72.8	5
June	134.8	7.5	342.5	13.6	56.9	9
July	240.2	96.5	493.2	24.1	39.1	14
August	210.0	65	374.5	21.1	34.0	13
September	133.7	8.8	463.5	13.4	73.4	8
October	22.8	0	210.4	2.3	162.9	2
November	6.2	0	66	0.6	202.5	1
December	20.3	0	107.7	2.0	127.8	1

**Rainy days were rounded.*

Table 1.
Mean, monthly rainfall characterization (mm) of month along with rainy days and CV (1964–2018).

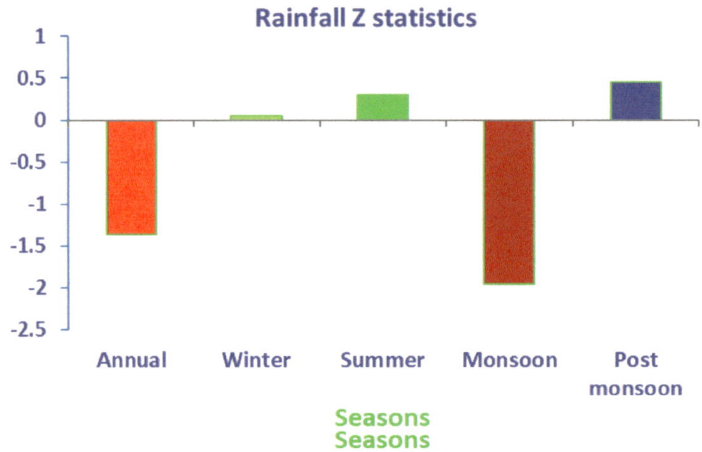

Figure 1.
Mann- Kendall test statistics for annual and rainfall data of different seasons.

2.1.4 Winter (December to February) and summer (March to May) rainfall

The out of 55 years, winter season rainfall was 30 years above the normal rainfall and rest years it was below the normal rainfall (**Figure 1**). The winter season rainfall have recorded slightly increasing trend. However, winter rainy days showing decreasing trend. The characteristics of the winter rainfall are mean rainfall (111.2 mm), highest rainfall (216.5 mm) in the year 2013, and lowest rainfall (13.0 mm) in the year 2016. The highest rainfall in single day (77.0 mm) on 18th February 2003] Likewise, mean winter rainy days (8.3 days); Highest rainy days in a year (15 days) in the year 1975; Lowest rainy days (2 days) in years 1964, 1967 and 2006}]. The summer rainfall in 22 years was above the normal and rest years, was below the normal (**Figure 1**). The summer rainfall and rainy days showed increasing trend (**Figure 1**). The mean rainfall was 136.1 mm, highest rainfall (369.5 mm) in 1983, lowest rainfall (29.0 mm) in 2013 and highest rainfall in single day (57.5 mm) on 7th May 1998. Similarly, mean annual rainy days were 12.4 days, highest rainy days in a year (23 days) in 1990 and lowest rainy days (4 days) in the years 1968, 1984 and 1992.

2.1.5 Monsoon (June to September) and post-monsoon (October to November) rainfall

The monsoon rainfall out of 55 years, 30 years was below the normal rainfall and it showed decreasing trend (**Figure 1**). The similar trend was observed with regard to rainy days. The mean rainfall was 718.8 mm, highest rainfall (1156.5 mm) in 2010, lowest rainfall (424.2 mm) in 2015 and highest rainfall in single day (167 mm) on 18th September 2010. Whereas, mean annual rainy days (45 days), highest Rainy days (60 days) in a years 1977,1988 and 2010 and lowest rainy days (27 days) in year 2009. The post monsoon rainfall in 36 years was below the normal rainfall and showed an increasing trend (**Figure 1**). The characteristics of post-monsoon rainfall exhibited mean rainfall (29.9 mm), highest rainfall (210.4 mm) in the year 1985, lowest rainfall (0.0 mm) in the years 1964, 1974, 1993, 1994, 2001 and 2017 and highest rainfall in single day (99.8 mm) on 13th October 1985. The mean annual rainy days were 2.0 days, highest rainy days in a year (11 days) in the year 1997 and lowest rainy days in year (0 days) in the years 1964, 1967, 1969, 1974, 1975, 1984, 1988, 1993, 1994, 2001 and 2017.

3. Estimation of crop water requirement in study area

Water requirement of the crops defines the quantity of water needed to meet the water losses through evapotranspiration of a disease-free crop under non restricting soil conditions, including soil, water and fertility and achieving the full potential under a given soil environment in a given time. Water requirement of the main *Kharif* and *Rabi* season crops (*i.e.* Vegetable pea, Barley, Rajma, Tomato, French Bean, Chili, Rice, Wheat, Maize, Soyabean, Okra, Mustard, and Cow pea) were calculated on the basis of Reference crop evapotranspiration (ET_o) on the monthly basis by using CROPWAT model based on FAO-Penman-Monteith's semi-empirical equation. The required weather data (min. and max. Temperature, rainfall, sunshine hours, wind speed, etc.) was collected from the automatic weather station. Reference evapotranspiration (ET_o) expresses the evaporative index of the atmosphere at a specific location. It is independent of crop type, stage of development and management practices. The reference evapotranspiration had been calculated using this Equation [21] in the following form:

$$ET_o = \frac{0.408\Delta(R_n - G) + \gamma \frac{900}{T+273} u_2(e_s - e_a)}{\Delta + \gamma(1 + 0.34u_2)} \tag{1}$$

Where,
ET_o = reference evapotranspiration [mm day^{-1}].
u_2 = wind speed at 2 m height [m s^{-1}].
R_n = net radiation at crop surface [MJ m^{-2} day^{-1}].
G = soil heat flux density [MJ m^{-2} day^{-1}].
T = mean daily air temperature at 2 m height [°C].
γ = psychrometric constant [kPa $°C^{-1}$].
Δ = slope of vapor pressure curve [kPa $°C^{-1}$].
$e_s - e_a$ = saturation vapor pressure deficit [kPa].

3.1 Crop coefficient

The crop evapotranspiration, (ET_c) is calculated by multiplying the reference crop evapotranspiration, (ET_o) by crop coefficient (K_c). Consequently, different crops will have different crop coefficients. K_c value varies with the type of crop, climate, soil evaporation and crop growth stages [21].

$$ET_c = K_c \times ET_o \tag{2}$$

Where,
ET_c = crop evapotranspiration [mm day^{-1}].
K_c = crop coefficient [dimensionless].
ET_o = reference crop evapotranspiration [mm day^{-1}].

3.2 Gross irrigation requirement

The gross irrigation requirement (GIR) accounts for losses of water included during conveyance and application of irrigation water to the field. The gross irrigation requirement is calculated by the ratio of net irrigation requirement to the irrigation efficiency.

3.3 Irrigation efficiency

Irrigation efficiency is defined as the ratio of amount of water beneficially used by plant as evapotranspiration to the amount of water applied to the plant area. The irrigation efficiencies under different methods *i.e.* 40%, 50%, 55%, 75%, and 90% are taken for border, check basin, furrow, sprinkler, and drip irrigation system, respectively [22].

3.4 Irrigation water requirement

The irrigation water requirement represents the difference between the crop water requirement and effective rainfall. Other factors or losses have minimal effect on irrigation water requirement and can be neglected [23] as shown in the equation below:

$$IR = ET_c - (P_e + G_e + W_b) \quad (3)$$

Where,
IR = irrigation requirement (mm).
Et_c = total crop evapotranspiration (mm).
P_e = effective rainfall (mm).
G_e = groundwater contribution from water table (mm).
W_b = water stored in the soil at the beginning of each period (mm).

3.5 Effective rainfall (P_e)

It is only a part of the rainfall that can be effectively used by the crop, depending on its root zone depth and the soil storage capacity. Total rainfall amount is not considered as effective rainfall; some part of rainfall may be lost through surface runoff, deep percolation or evaporation. FAO CROPWAT ver. 8.0 model could be used rainfall data and employs the USDA S.C. method approach to estimate effective rainfall on a daily basis or monthly basis.

4. Irrigation scheduling

Irrigation scheduling is the process of deciding the period and quantity of irrigation water during the crop growth under different irrigation methods [24]. Its main objective is to apply irrigation at the right period and in right amount. Irrigation amount is determined in terms of gross irrigation requirement and pumping time per application, while, irrigation time is based on depletion of soil moisture content of the crop root zone reached at critical point [25]. That is basically dependent on the consumptive use rate of crop and method of water application to the plant root zone [26]. The quantity of irrigation water for each treatment was calculated based on the soil moisture content before irrigation and root zone depth of the plant using the Eq. 1.4:

$$SMD = (\theta_{FC} - \theta_I) \times D \times Bd \times MAD \quad (4)$$

Where,
SMD = Soil moisture deficit (mm).
θ_{FC} = Soil moisture content at field capacity (%).
θ_I = Soil water content before irrigation (%).

D = Depth of root development (mm).
Bd = Bulk density of the particular soil layer (g cm^{-3}).
MAD = Management allowable Depletion (%).

4.1 Management allowable depletion (MAD)

Producing optimal yield requires that the soil water content be maintained between an upper limit at which leaching becomes excessive and a lower point at which crops are stressed [27]. As water is removed from the soil through ET, there is a point below which the plant experiences increasing water stress. This point is known as the management allowable depletion (*MAD*). The typical MAD values considered are 33% for shallow-rooted, high value crops; 50% for medium-rooted, moderate value crops and 67% for deep-rooted, low value crops [28]. Selection of MAD value for different crops with respect to soil type, initial field capacity (FC), permanent wilting point (PWP), and threshold soil moisture content (TSMC) must be determined. Threshold soil moisture content ascertains what fraction of soil is allowed to dry before the next irrigation event. Threshold soil moisture content can be determined in the following form:

$$\theta_{TSMC} = \theta_{FC} - MAD\ (\theta_{FC} - \theta_{PWP}) \quad (5)$$

Where,
θ_{TSMC}: Soil moisture content at threshold level (%).
θ_{FC}: Soil moisture content at field capacity (%).
θ_{PWP}: Soil moisture content at permanent wilting point (%).

The determination of soil moisture content at threshold level is most important factor for irrigation scheduling on real time basis. This value varies with crop, soil, climate and crop growth stages. Whenever the soil moisture content at field capacity is depleted through ET, percolation losses, etc. to equal or below the θ_{TSMC} value, irrigation scheduling must be given otherwise crop yield and plant growth will be affect harmful way.

5. Estimation of water requirement of major crops using CROPWAT model

CROPWAT is a decision support tool developed by the land and water development division of FAO. CROPWAT model is extensively tested, widely accepted for calculation of crop water requirements based on soil, climate and crop data. In addition, the program allows the development of irrigation schedules for different management and the calculation of scheme water supply for varying crop patterns. CROPWAT ver. 8.0 model used weather data and employs the modified penman-monteith approach used to estimate reference evapotranspiration on a daily basis. The meteorological data was taken from Agromet Observatory, ICAR-VPKAS, Experimental farm Hawalbagh, Almora. The mean annual rainfall at experimental site was 1000.13 mm. The general soil properties of the experimental field were used in CROPWAT model. Based on the details of soil characteristics, total available water was taken 135 mm m^{-1} depth of soil. Infiltration rate was measured using double ring infiltrometer and the basic rate was 6.8 mm $hr.^{-1}$ and the unsaturated hydraulic conductivity was 0.77 cm h^{-1}. The estimation of irrigation water requirement (346–376 mm, 131–189 mm, 1.4 mm, 1.3 mm, 78.6 mm, 93.5 mm 104.1 mm, 176 mm, 96.9 mm, 16.2 mm, 18.3 mm, 15.5 mm, and 7.4 mm of Rice, Wheat, Maize, Soybean, Vegetable Pea, Rajma, Barley, Tomato, French Bean, chili, okra, mustard

Crop name	Crop water requirement (mm)	Effective rainfall (mm)	Irrigation water requirement (mm)
Vegetable pea	209.8	131.7	78.6
Barley	250	147.9	104.1
Rajma	123.3	29.7	93.5
Tomato	262.4	84.3	176
French Bean	166.9	70.1	96.9
Chili	310.6	370.7	16.2
Rice	434–505.9	464–491	346–376
Wheat	269.2–375.1	163–194.6	131–189
Maize	247.9	395	1.4
Soybean	350.6	494.4	1.3
Okra	234.9	470.1	18.3
Mustard	155	145.5	15.5
Cowpea	171.9	280.2	7.4

Table 2.
Estimation of irrigation water requirement of major crops grown on experimental site using CROPWAT model.

and cowpea crop, respectively) and irrigation schedule plan of major crop was calculated using CROPWAT Model as presented in **Table 2**.

5.1 Water budgeting equation

This equation could be used to measure evaporation, seepage from pond and volume of water available in pond. The water budget method of determining long term available water present in pond can be used as a standard for comparing other methods. This method is not most accurate, but could be used satisfactory for practical purpose. The volume of water available in pond can be calculated using this equation in the following form:

$$\sum_{i=1}^{12} P_p + \sum_{i=1}^{12} R_{cbt} - \left[\sum_{i=1}^{12} S_p + \sum_{i=1}^{12} Ev_p + \alpha \sum_{i=1}^{12} \sum_{j=1}^{n=crop} WR_{crop} \right] = \sum_{i=1}^{12} WS_p \qquad (6)$$

Where,
P_p = precipitation in surface area of pond (m^3),
R_{cbt} = runoff from conservation bench terraces areas or plain surface (m^3).
S_P = seepage losses from Pond (m^3),
Evp = evaporation losses from Pond (m^3).
W_{Rcrop} = water requirement of crop (m^3).
WS_p = water storage in pond (m^3).

6. Ways to increase carbon storage in tree based land use systems

Carbon sequestration can be enhanced by adopting plantations or agroforestry. Loss of carbon storage can be prevented by reducing felling of forests, blocking or reducing emissions from agricultural activities and by reduced use of energy, oil

Agroforestry/land use systems	Age	Average vegetation C (Mg ha^{-1} y^{-1})
1. Fodder bank, Segu, Mali,South Africa, Sahel	7.5	0.29
2. Live fense, Segu, Mali, South Africa, Sahel	8.0	0.59
3. Tree based intercropping, Canada	13.0	0.83
4. Park lands, Segu, Mali,South Africa, Sahel	35.0	1.09
5. Agrisilviculture, Chhatisgarh, Central India	5.0	1.26
6. Silvopasture, South Oreogaon, USA	11.0	1.11
7. Silvopastoralism, Kurukshetra, India	6.0	1.37
8. Silvopastoralism, Kerala, India	5.0	6.55
9. Cocoa agroforestry, Makoe, Cameroon	26.0	5.85
10. Cocoa agroforestry, Durialban, Costarica	10.0	11.08
11. Shaded coffee, South-West Congo	13.0	6.31
12. Agroforestry woodlots, Partorico	4.0	12.04
13. Agroforestry woodlots, Kerala, India	8.8	6.53
14. Home and farm garden	23.2	4.29
15. Indonesian homegarden, Sumatra	13.4	8.00
16. Mixed species stand, Puertorico	4.0	15.21

Source: [29].

Table 3.
Carbon sequestration potential in vegetation of world's major agroforestry systems.

and fertilizer. It is the biggest practical alternative, low-cost as well as associated ecological advantages, compatibility to reduce poverty, role of the social dimension/expansion are the key to global climate change mitigation and adaptation [7, 9, 12]. Carbon storage in agroforestry remains in aboveground biomass (wood biomass), leaf group (foliage), shrub, vine, herb, dead biomass (dead wood, litter) and below ground biomass (roots), soil organic carbon etc. According to Nair et al. [29], the world's vegetation carbon sequestration ability by major agroforestry systems is listed in **Table 3**.

7. Status of agroforestry in Himalaya

In Himalayan region of India agroforestry is promising and distributed in large area in different forms. The area ranged from 4.95 in a watershed to 137 ha in of North-west Himalaya (**Table 4**). Carbon storage ranged from 3.31 to 31.71 t/ha in Indian Himalaya (**Table 5**).

System	Area (ha)	Region	Author
Cardamom agroforestry	27.59	North-East Himalaya	[30]
Agroforestry	4.95	North-West Himalayan Watershed	[31]
Willow based agroforestry	137	North-West Himalaya	[20]

Table 4.
Reported area under agroforestry systems of Himalaya.

Agroforestry system	C Storage (t/ha)	Region	Author
Silvopasture	31.71	Himachal Pradesh	[32]
Agrisilviculture	13.37		
Agrihorticulture	12.28		
Agrisilviculture	15.91	Uttarakhand	[33]
Agrihorticulture	12.15	Himachal Pradesh	[31]
Agrisilviculture	12.02	Uttarkhand	[34]
Silviculture	4.4	Uttarakhand	[35]
Silviculture	3.31–3.95	North-East Himalaya	[36]

Table 5.
Carbon storage of different agroforestry systems of Himalaya.

7.1 Oak high-density plantation

The experiments were conducted on high-density plantation of oak (*Quercus leucotrichophora*) for proper management of tree canopy with four lopping techniques. The lopping techniques included: pollarding at 1 and 2 meters (backwards cutting of the tree trunk so that the dense numbers of branches can be generated); Local practices (slightly above from where the branches split leaves and tender twigs removed in random manner); without disturbing the upper 1/3 part of the tree lower 2/3 part pruned for fodder leaves (lopping) (**Table 6**). It was found that at the age of 30 years oak tree can store 86.7 to 356.9 Mg ha^{-1} carbon stock, 317.2 to 1306.5 Mg ha^{-1} biomass carbon dioxide and carbon sequestration in the range of 2.9 to 11.9 per Mg per hectare per year were found in various cutting management [16].

7.2 Fruit based agrihorticulture

In the fruit based agrihorticulture system the highest carbon stock was found in pear + wheat (17.0 Mg ha^{-1}) followed by apricot + wheat, plum + wheat, hill lemon + wheat and wheat with 11.9, 10.0, 8.4 and 4.8 Mg ha^{-1} respectively, (**Figure 2**). Similarly, biomass carbon dioxide was 62.3, 43.6, 36.5, 30.9 17.6 Mg ha^{-1} in pear + wheat, apricot + wheat, plum + wheat, hill lemon + wheat and wheat, respectively [14]. Carbon sequestration in the range of 4.7 to 5.3 Megagarams carbon per hectare per year was found in different treatment.

Treatment	Carbon stock (Mg ha^{-1})	C sequestration (Mg ha^{-1} yr.$^{-1}$)	Biomass CO_2 (Mg ha^{-1})
Coppicing at 1 m	86.7	2.9	317.2
Local	169.6	5.7	620.6
Pollarding at 2 m	123.4	4.1	451.5
1/3rd top portion undisturbed	356.9	11.9	1306.5
Source: [16].			

Table 6.
Effect of different cutting management on carbon stock, carbon sequestration and biomass carbon dioxide.

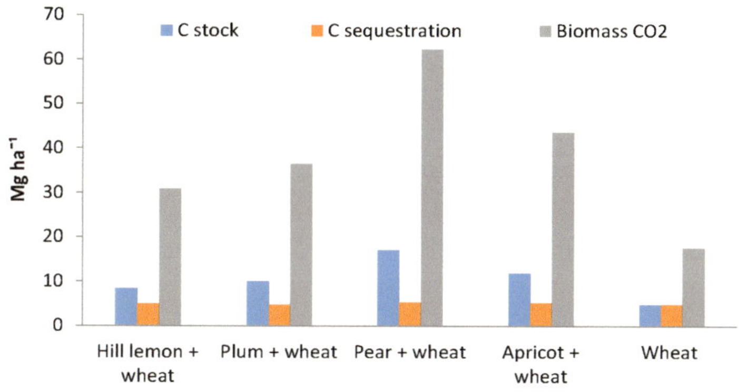

Figure 2.
Aboveground carbon stocks, carbon sequestration and biomass CO_2 in fruit tree based land use systems [8].

7.3 Agrihorticulture

Pecan nut (*Carya illinoinensis*) based agrihorticulture system in which pecan nut + lentil, pecan nut + wheat, lentil and wheat were grown (**Table 7**). Carbon stock of 23.9 and 25.3 Mg ha^{-1} with lentil and wheat and biomass carbon dioxide 92.85 and 87.78 Mg ha^{-1} with wheat and lentil was recorded, respectively [11]. In peach (*Prunus persica*) biomass C stock was recorded 19.4 Mg/ha under agrihorti system.

Therefore, in the context of climate change and to meet the need of rapidly growing population agroforestry is very important. Agroforestry is very essential for pollution free atmosphere and for feed as well as fuel, timber and the unemployed youth in the village could get jobs through agroforestry.

7.4 Silvipasture

In Silvipastoral system among trees oak (*Quercus lecotrichophora*) stored significantly high carbon (24.85 t/ha) as compared to rest trees (**Figure 3**). Whereas, among cutting management pollarding at 3 m height stored highest carbon stock (14.87 t/ha) than other cutting strategies.

7.5 Terrace and wayside plantation

Carbon stock was measured in linear strip plantation of Kachnar (*Bauhinia retusa*). In Kachnar terrace plantation highest @ 3.17 ± 0.88 t/1000 m length, carbon was stored in lopping of lower ½ (half) part and keeping top ½ parts undisturbed of the trees (**Figure 4**). However highest carbon stock of 2.60 ± 0.32 t/

Treatment	Carbon stock (Mg ha^{-1})	Biomass CO_2 (Mg ha^{-1})
Pecan nut + Lentil	23.92	87.78
Pecan nut + Wheat	25.30	92.85
Lentil	1.17	4.29
Wheat	2.50	9.17

Source: [11].

Table 7.
Carbon stock and biomass CO_2 in pecan nut based agrihorticulture system.

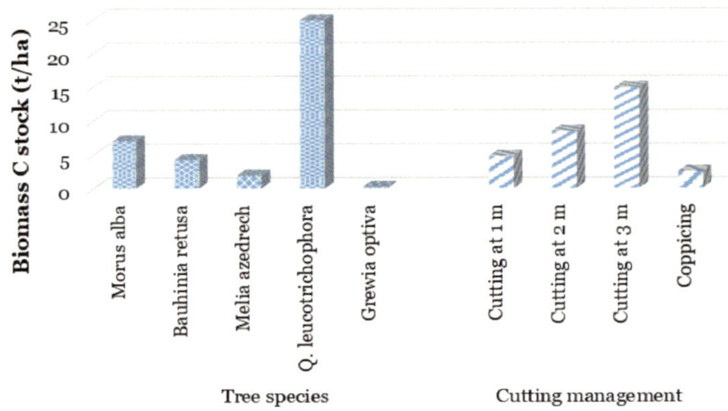

Figure 3.
Biomass carbon (C) stock in silvipastoral system on marginal land.

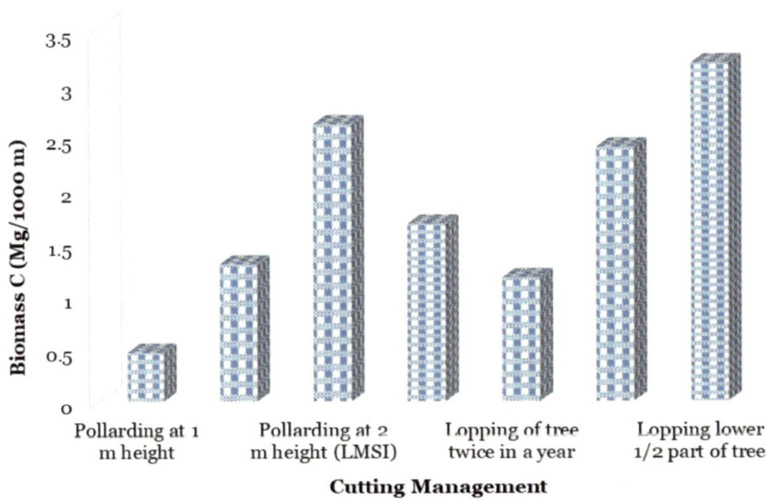

Figure 4.
Biomass C in terrace plantation of Kachnar cutting management in terrace plantation.

1000 m length, was recorded in lopping of lower 2/3 part and keeping top 1/3 undisturbed of Kachnar in wayside plantation (**Figure** 5).

7.6 Silvihorticulture

In silvihorticulture system, significantly high carbon stock (281.6 t/ha) was recorded in Kharik (*Celtis australis*) as shown in **Figure 6** followed by in oak (*Quercus leucotrichophora*), Kachnar (*Bauhinia retusa*) and least in Bhimal (*Grewia optiva*).

8. Conclusion

We conclude that the wheat yield decline due to higher maximum temperature and rice due lower temperature. Therefore, it is essential to know the effects of different weather factors on production and productivity. In view of climate change, it is also demand of the time to explore alternative of crop production system to reduce climate impact of climate change and its variation on agriculture

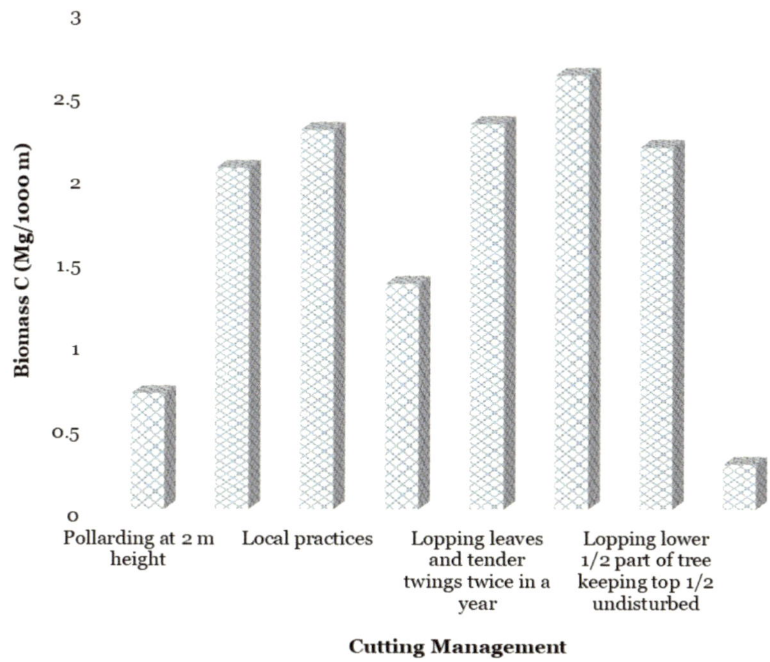

Figure 5.
Biomass C stock in wayside plantation with different cutting management in Kachnar in wayside plantation.

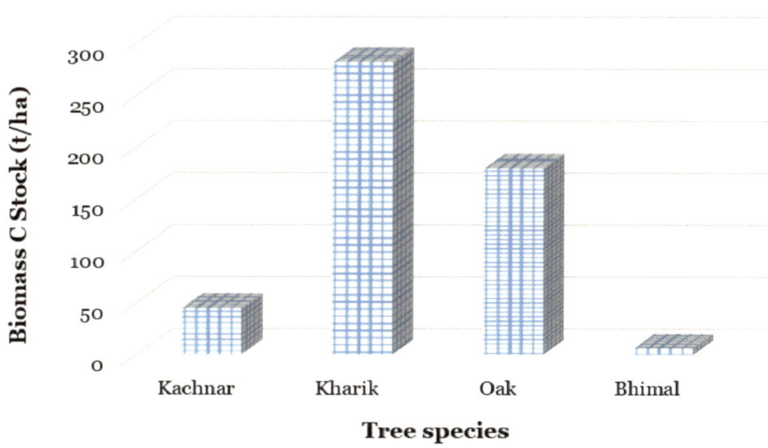

Figure 6.
Biomass C stock in different trees in silvihorticulture system.

and to maintain productivity of resources and also maintain and improve the health of resources. The inclusion of tree i.e. forest or horticulture along with crop will tremendously help in reducing climate change and variation effects on production system. Trees convert poisonous gas CO_2 into lifesaving oxygen (O_2). This action purify the air as well as help in preventing elevated temperature because trees absorb atmospheric carbon in the process of photosynthesis. In agroforestry, we should choose trees with short life cycle and fast growing in nature. India has diversified climate, different agricultural conditions and abundant wasteland in villages need to be considered while implementing various methods of forestry. Hence, to take along sustainability proper considerate of climate is necessary in hill agriculture through crop sowing window adjustment and other processes as per climate appropriateness.

Acknowledgements

The authors are grateful to the Director, Indian Council of Agricultural Research- Vivekananda Parvatiya Krishi Anusandhan Sansthan, Almora, Uttarakhand for providing necessary support.

Conflict of interest

The authors hereby declare that there is no conflict of interest.

Author details

Ram Prakash Yadav*, Suresh Chandra Panday, Jitendra Kumar, Jaideep Kumar Bisht, Vijay Singh Meena, Mahipal Choudhary, Shyam Nath, Manoj Parihar and Rajendra Prasad Meena
ICAR-Vivekananda Parvatiya Krishi Anusandhan Sansthan, Almora, Uttarakhand, India

*Address all correspondence to: rams34052@gmail.com

References

[1] Ramakrishna YS, Rao GGSN, Kesava Rao AVR, Vijaya Kumar P. Weather resource management. In: Natural Resource Management for Agricultural Production in India. (Yadav JSP and Singh GB(eds) International Conference on Managing Natural Resources for Sustainable Agricultural Production in the 21st Century, New Delhi, India 2000;245–370 pp.

[2] Samra JS. Review and analysis of drought monitoring, declaration and Management in India. IWMI, 2004, working paper 84. Colombo. Srilanka: International Water Management Institute.

[3] Howden M, Soussana JF, Tubeiello FN, Chhetri N, Dunlop M, Meinke H. Adapting agriculture to climate change. Proceedings of the National Academy of sciences (PNAS). 2007;**104**:1969119696

[4] Rupa Kumar, K, Sahai AK, Krishna Kumar K, Patwardhan SK, Mishra PK, Rewadekar K, Kamala, Pant GB. High resolution climate change scenarios for India for the 21st century. Current Science 2006;**90**:334–345.

[5] Panday SC, Srivastva AK, Mahanta D, Mina BL, Agrawal PK, Bhattarcharya R, Singh KP, Kumar A, Bhatt JC. Imapct assessment of climate change on rice and wheat grain yield for the Himalyan region of Uttrakhnad. In extended summaries in Internation at Conference Preparing agriculture for climate change February 6–8 2011 held at PAU Ludhiana 2011;pp 43.

[6] Panday SC, Srivastva AK, Murty NS, Bisht JK, Bhatt JC, Gupta HS, Pattanayak A. Climate and Crop Planning in Mid Hills of Uttarakhand. ICAR-VPKAS Almora. 2016;Pp176.

[7] Yadav RP, Bisht JK, Gupta B, Mondal, T. Toward the C sequestration potential of agroforestry practices to combat climate change in Kumaon Himalaya, India.In: Bisht JK, Meena VS, Mishra PK, Pattanayak A, Editors. Conservation Agriculture. Springer. 2016; p. 293–313. DOI: 10.1007/978-981-10-2558-7_11.

[8] Pattanayak A, Bisht JK, Yadav RP, Pandey BM. Fodder tree based agroforestry systems in hills of Uttarakhand. In: Dev I, Ram A, Kumar N, Singh R, Kumar D, Uthappa AR, Handa AK, Chuaturvedi OP, editors. Agroforestry for Climate Resilience and Rural Livelihood. Scientific. New Delhi, India: Publishers; 2018. pp. 65-81

[9] Yadav RP, Gupta B, Bhutia PL, Bisht JK. Socioeconomics and sources of livelihood security in central Himalaya, India: A case study. International Journal of Sustainable Development & World Ecology. 2017;**24**(6):545-553

[10] Yadav RP, Bisht JK. Agroforestry: A way to conserve MPTs in North Western Himalaya. Research journal of agriculture and forestry. Science. 2013; **1**(9):8-13

[11] Yadav RP, Bisht JK, Bhatt JC. Biomass, carbon stock under different production systems in the mid hills of Indian Himalaya. Tropical Ecology. 2017;**58**(1):15-21

[12] Yadav RP, Bisht JK, Meena VS, Choudhary M. Sustainable Agroecosystems for Livelihood Security in Indian Himalaya. DeOliveira AB, edtor. Sustainability of Agroecosystem. IntechOpen: In; 2018. pp. 63-77. DOI: org/10.5772/intechopen.74495

[13] Yadav RP, Bisht JK. Agroforestry to combat climate change. Kheti. 2014; **66**(10):11-14

[14] Yadav RP, Bisht JK, Pandey BM. Aboveground biomass and carbon stock

of fruit tree based land use systems in Indian Himalaya. The Ecoscan. 2015;**9** (3&4):779-783

[15] Yadav RP, Gupta B, Bhutia PL, Bisht JK, Pattanayak A. Biomass and carbon budgeting of land use types along elevation gradient in Central Himalayas. Journal of Cleaner Production. 2019;**211**:1284-1298. DOI: 10.1016/j.jclepro.2018.11.278

[16] Yadav RP, Bisht JK, Pandey BM, Kumar A, Pattanayak A. Cutting management versus biomass and carbon stock of oak under high density plantation in central Himalaya, India. Applied Ecology & Environmental Research. 2016;**14**(3):207-214

[17] IPCC (Intergovernmental Panel on Climate Change). Good practice guidance for land use, land use change and forestry. IPCC national greenhouse gas inventories programme. Hayama; Japan. 2003:295

[18] Yadav RP, Gupta B, Bhutia PL, Bisht JK, Pattanayak A. Sustainable agroforestry systems and their structural components as livelihood options along elevation gradient in central Himalaya. In: Biological Agriculture & Horticulture. 2018. DOI: org/10.1080/01448765.2018.1457982

[19] Yadav RP, Gupta B, Bhutia PL, Bisht JK. Sustainable agroforestry systems for livelihood security and their economic appraisal in Indian Himalayas. Economic Affairs. 2018;**63**(3):01-07

[20] Chavan SB, Keerthika A, Dhyani SK, Handa AK, Newaj R, Rajarajan K. National Agroforestry Policy in India: A low hanging fruit. Current Science. 2015; **108**(10):1826-1834

[21] Allen RG, Pereira LS, Raes D, Smith M. Crop evapotranspiration-guidelines for computing crop water requirements. FAO irrigation and drainage paper no. 56 Rome. Italy.

[22] Sivanappan RK, Padmakumari O, Kumar V. Drip Irrigation, **Pp** 412. Keerthi Publishing House Pvt Ltd. Tamil Nadu: Coimbatore; 1998

[23] Savva AP, Frenken K. Crop water requirements and irrigation scheduling. Irrigation Manual Module. 2002;**4**:122

[24] Jensen MJ. Summary and challenges. 1981;pp. 225-231. In proc. of the ASAE's irrigation scheduling Conf., irrigation scheduling for Water & Energy Conservation in the 80's, ASAE Publ. 23-81, Am. Soc. Agric. Engr., St. Joseph. MI.

[25] Ramadan MH, El-Adl MA, Abd El-Mageed HN, Maher M. Computer-aided mapping irrigation scheduling for Arab Republic of Egypt. *The 2nd International Conference on Water Resources & Arid Environment.* 2006;agriculture, Mansoura University. Egypt.

[26] Pal MS. Recent Advance in Irrigation Water Management. India: Kalyani Publishers Ludhiana; 2012. pp. 409-438

[27] Keller J, Blienser RD. *Sprinkler and Trickle Irrigation*. 1990;pp 19-95. Van Nostrand Reinhold. New York.

[28] Cuenca RH. *Irrigation System Design-An Engineering Approach,* prentice-hall, Inc., Englewood, cliffs, NJ, 1989;pp 552. USA.

[29] Nair PKR, Nair VD, Kumar BM, Showalter JM. Carbon sequestration in agroforestry systems. Advances in Agronomy. 2009;**108**:237-307

[30] Sharma R, XU J, Sharma G. Traditional agroforestry in the eastern Himalayan region: Land management system supporting ecosystem services. Tropical Ecology. 2007;**48**(2):1-12

[31] Goswami S, Verma KS, Kaushal R. Biomass and carbon sequestration in different agroforestry systems of a

Western Himalayan watershed. Biological Agriculture & Horticulture. 2014;**30**(2):88-96

[32] Verma KS, Kumar S, Bhardwaj DR. Soil organic carbon stocks and carbon sequestration potential of agroforestry systems in H.P. Himalaya region of India. Journal of tree. Sciences. 2008; **7**(1):14-27

[33] Kaushal R, Tewari SK, Banik RL, Chaturvedi S. Growth, Biomass production and soil properties under different bamboo species. In: ISTS-IUFRO Conference on Sustainable Resource Management for Climate Change Mitigation and Social Security at Chandigarh, India. (March 13–14, 2014).

[34] Singh P, Lodhiyal LS. Biomass and carbon allocation in 8-year-old poplar (*Populus deltoides* marsh) plantation in Tarai agroforestry systems of central Himalaya. India. New York Science Journal. 2009;**2**:49-53

[35] Dhyani SK, Puri DN, Narain P. Biomass production and rooting behaviour of Eucalyptus tereticornis Sm. On deep soils and riverbed bouldery lands of Doon Valley, India. Indian Forester. 1996;**122**(2):128-136

[36] Negi JDS, Bahuguna VK, Sharma DC. Biomass production and distribution of nutrients in 2 years old teak (Tectonagrandis) and gamar (Gmelinaarborea) plantation in Tripura. Indian Forester. 1990;**116**(9):681-686

Chapter 3

Climate as the Major Factor Controlling Phenology

Boubakeur Guesmi

Abstract

The witnessed aberrance and irregularities in the timing of pheno-phases is an undeniable evidence of the reality of the climate change and hence proves the complete control of climate over phenology. In fact, some researchers mentioned the advance of blooming and the delay of defoliation to the mid of winter as well as the disappearance of many animal and vegetal species. This would visibly illustrates the impact of climate changes which became a factual reality. These facts a long with the rhythmicity of life under the climate control and seasonality makes the importance of this chapter unequivocal, and a backbone for this very book of "Agrometeorology". Accordingly, this chapter treats each phenophase from dormancy to fructification to cover all the plant life cycle. For each of which we focus on how climate is intimately controlling the biological processes of each life phase and how climatic elements are the strongest and first factor which induces plant to starts the appropriate phenophase according to the fitting season. Plant is indeed very sensitive to seasonal variation in climate elements which induces the transcription of specific genes to produce specific enzymes which to their turn are with specific act on specific cells and tissues. Hence there is a high harmony between plant physiological response and climate seasonality endorsed by the circadian clock which is merely created by the historical subjection of plants to the impact of climate. Nonetheless, the recent climate changes are seemingly to be against this natural harmony between phenology and climate. This should due to their erraticism which may cause damages to the ecosystem and available resources. Thence, this chapter within this book would be inspiring for some strategies of adaptation to the climate changes to avoid such a prejudice against crops by adjusting the agricultural calendar and planting dates to avoid coincidence of fragile phenophases (germination, flowering, and fructification) with climatic hazards.

Keywords: phenophase, climate change, agriculture, germination, fructification, food security

1. Introduction

The autumn is usually called fall due to the featuring fall of leaves during this season as a phenological phenomenon called defoliation. This would figure out how intimately related are seasons and phenology or in other words how strong is the repercussion of climate seasonal variation on the plants and animals successive and cyclic phenomena. Indeed, we almost notice seasons via phenological variation along with climatic one. This deep intimate co-occurrence between climate and phenology phenomena is very appealing for wonder and consequently to the exploration of the relations and mechanisms of these relations and how living beings respond to climate seasonal variations during their life cycles.

As a matter of undeniable fact, this chapter consists of a core and mandatory knowledge for the agriculture and farmers to make their crops safe and the most productive. Therefore, this chapter is likely to be the backbone for this very book of agro- meteorology and all disciplines within the same scope and the nature and life's scope.

Particularly, in the recent context of climate change. Indeed, Subrahmanyam and Murthy (in Ref. [1]) confirmed that global climate change is a reality, a continuous process that needs to be taken seriously. Many evidences have been gathered to depict that climate change is taking place. And although several species have responded to climatic changes throughout their evolutionary history, there is a concern as to how different ecosystems and populations will respond to this rapid rate of change.

The fact of the book being devoted to agro-meteorology imposes that this chapter should be more specified and oriented towards phenology of plants as to they are the subject of agriculture more than animals. This is also due to another fact that climate impacts plants more than animals. Accordingly, this chapter attempts to explain how climate factors affect and control each pheno-phase within the plant life cycle from dormancy and germination to fructification, after defining all of phenology, phonological phases, climate, climate factors, and climate seasonal variations.

2. Phenology

In general, Phenology is the observation and measurement of events in time [2]. Obviously, the word phenology is composed of the part "pheno" which refers to phenomena and "logy" which is commonly known to mean science or study, thus phenology is literally the study or science of phenomena. Practically, Phenology refers to the study the cyclic phenomena of living beings mainly plants and animals including insect and their succession in seasons as well as their timings [1, 3]. This is under the direct influence and control of climate and surrounding environmental conditions including the duration of sunlight, precipitation, temperature and other life-controlling factors [1]. The recent climate changes makes the deep mastery and comprehension of phenology worthwhile, because some unpredicted climate changes could cause many crop damages. Actually, the assessment of impacts of projected climate changes on natural ecosystems is not based on accurate scientific modeling or field studies at regional level [1], and It is well documented that plant and animal phenology is changing in response to recent climate warming in the Palearctic [3]; besides, global change, encompassing natural and anthropogenic changes to the Earth system at sub-annual to geologic time scales, has strong interactions with vegetation phenology [4]. Thus phenology is central and crucial as a background for the discipline of agronomy and mainly agro-meteorology to predict the eventual response of living being to the unpredictable climate changes and therefore probable agricultural damages.

The term of phenology was first introduced in 1853 by the Belgian botanist Charles Morren. It refers to the science that measures the timing of life cycle events for plants, animals, and microbes, and detects how the environment influences the timing of those events. Namely, it focuses on how environmental factors mainly climatic variables influence the phenol-phases to hence make a harmony between seasons and life cycle events including defoliation, plant dormancy, leaf budburst, blooming and first flower, last flower, first ripe fruit, and leaf shedding, and for animals this includes molting, mating, egg-laying or birthing, fledging, emergence from hibernation, and migration. Thus, phenologists record the dates when every event occurs, its duration and how environmental conditions such as temperature and precipitation affect its timing [2].

3. Climate and weather

The timing of phenological events can be quite sensitive to environmental conditions mainly climatic [2, 4, 5]. For example, an advance of leaf budburst and blooming for could be caused by warming an drought in spring and this could be for two weeks earlier than usual, whereas cold and moisture could exceptionally could equally delay them. Thus, weather and climate controls the timing of phenol-phases which vary among years [2]. Effectively, climate cyclic variations are the controlling variables of phenol-phases timing. In plants, bud-burst, leaf-expansion, abscission, flowering, fertilization, seedset, fruiting, seed dispersal and germination all take place in due season [6].

On the one hand weather is limited to a very short period of time from one day to less than week and it includes atmospheric conditions of a region, such as temperature, precipitation, humidity, wind, and sunshine. The climate of a region, on the other hand, concern a long period more than thirty years commonly defined as the conventional period for climatic studies. Climate consists of the generally-prevailing weather conditions for this period and in a large geographic region. For example, Santa Barbara, California is characterized by a Mediterranean climate – warm, dry summers and cool, moist winters. There are, however, daily and weekly changes in the weather that can rapidly change the temperature, sunshine, and wind conditions [2]. Nonetheless, there is no steady rhythm for all years particularly in the context of recent climate changes. This would not be tolerated by some species and therefore they are likely been extinguished and rarified in many regions.

4. Climate elements controlling phenology

Temperature, solar radiation, and water availability are assumed to be the key factors that control plant phenology [4]. However, not only these climatic factors. Indeed, temperature is an inevitable factor on which depends all the chemical reaction and mainly those occurring inside cells of living organism. In addition the temperature is both a characteristic of live and an indispensable condition to survive. Some biological functions and reaction may be inhibited or stopped by cold like in hibernation and dormancy. The solar radiation is unequivocally source of energy which is transformed from it luminous form to the chemical form (ATP) by the photosynthesis in the chlorophyll within plants. Furthermore, sun light is factor to fixate the calcium, to product vitamin and the duration of insolation which is called photoperiod determines the season of fall, season of bud bursting and blooming. Precipitations are source of water for crops. Water is indispensable for any form of life on earth. Indeed water is the solvent in all physiological solution in living organisms. As well as it transmits nutrients and regulates temperature of bodies of living beings.

In fact all the climate with its elements and their features including duration, frequency and intensity are influencing phenophases and living beings lives whence the elaboration of the discipline of bioclimatology.

5. Climate, overwintering and defoliation

In period of tough climatic conditions some plants and animals can no longer neither resist nor adapt to rude conditions of autumn and winter. Therefore, they adopt a specific strategy to survive. This is possible by pausing growth and development, which can occur in different organs like seeds and buds. This is known as dormancy which is controlled both by genetic and environmental factors. As the most studied dormancy, seed dormancy is an important adaptive trait in wild plants. The plant hormone abscisic acid plays a crucial role in the establishment and maintenance of dormancy, whereas gibberellins promote germination. The abscisic hormone (ABA) is specific to plants, and plays many roles for plant responses to stresses such as drought, salinity, cold and freezing tolerance, heat stress and heavy metal ion tolerance. Dormancy is a main determinant period in the plant life cycle. It has strong variation between species [7].

According to the predictability of climate, the dormancy may be preventive or consequential. Predictive dormancy is when plants can predict the onset of winter through the short photoperiod and the decrease of temperature, however when climate is unpredictable and has a sudden changes, the organism enter directly in consequential dormancy after adverse conditions. This last may cause a high rate of mortality before entering in consequential dormancy which is a protection strategy. Furthermore, the biological clock in many spices determine autonomously the period of the year for every phenophase.

In soil, seeds dormancy is continually adjusted by a set of environmental signals (**Figure 1**). The time of the year is determined by signals related to the slow seasonal change and this may indicate how sensitive the plant sensors mainly in seeds are. The figure illustrates the range of environmental signals and how they can potentially inform the seed of the time [8]. As buried and incorporated into soil, seeds responds to a wide range of edaphic and physical conditions to inform about the time of year and its appropriateness to the germination as illustrated in the **Figure 1**. The nitrate is commonly known to have a very important role in informing plants about the surrounding environmental conditions.

For the evergreen plants like some trees such as conifers, the dormancy consisted of the sustained light quenching for the whole winter period by the

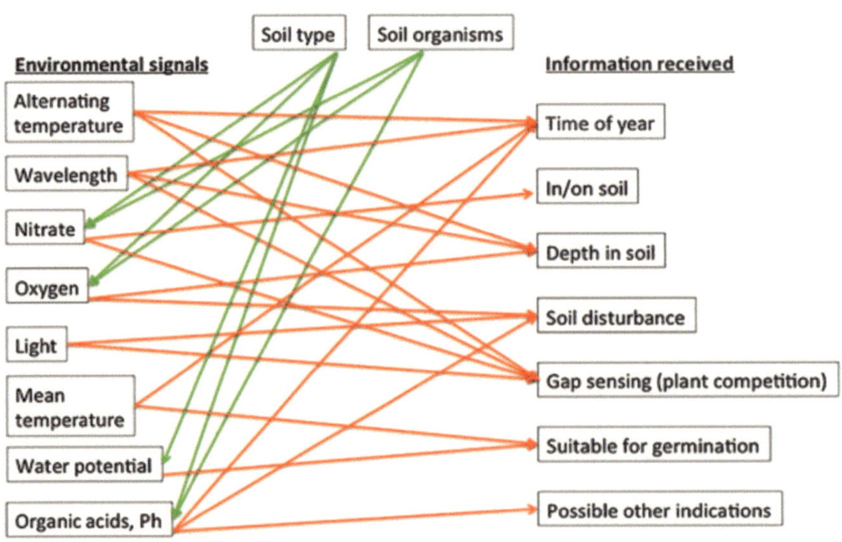

Figure 1.
Environmental signals controlling seeds dormancy and germination. ***Source:*** *Reference [8].*

xanthophyll-mediated non-photochemical antenna. This is a form of a protection for the evergreen foliage from photo-oxidative damage when photosynthesis is restricted or prevented by low temperatures during the winter. The molecular mechanisms of this cold acclimation are still unknown, it implies alterations in the photosystem II antenna. Photosystem I is also involved via its support of cyclic electron transport at low temperatures, and also by non-photochemical quenching of absorbed light irrespective of temperature. Processes like chloro-respiration and cyclic electron transport may also be important for maintaining the functional integrity of the photosynthetic apparatus of overwintering evergreens both during periods of thawing in winter and during recovery from winter stress in spring [9].

Defoliation (removal of leaves), on the other hand and for the deciduous plants, is the strategy to minimize or stop the photosynthesis in the overwintering period. Defoliation accelerates sink metabolism and hence remobilizes carbon and nitrogen reserves, leading to improved source-sink relations. Through removing lower and senescing leaves, plant can assure a greatest capacity of photosynthesis and carbon and nitrogen metabolism in despite of adverse conditions [10]. Hence, the defoliation consists of a balancing between the minimum of photosynthesis supplies in adverse conditions and the plant needs.

6. Climate, germination and dormancy break

Whenever conditions are favorable, dormancy and overwintering become useless. Plants and seeds start anew their active lives. This starting is accomplished through germination in seeds characterized by the emergence of embryo from seed enclosing covers the endosperm, perisperm, testa, or pericarp. This metabolism is mainly activated by the seeds imbibition by water. This would incite the respiration metabolism to provide the necessary energy for the expansion of the embryo and after that the radicle through the covering tissues of the seed. The emergence of the radicle out of the seed indicates the germination completed and hence called the visible germination which ends up with the seeds germinated. Germination does not include the seedling growth [11].

In fact, not only water is the climatic factor inducing germination, there is also temperature and sunlight. However, the most essential environmental factor required for seed germination is water. Water availability acts following a specific model, the hydro-time model of germination [11].

While water availability and imbibtion of seed are indispensable to launch germination, temperature is important as well for germination and for all physiological functions both for animal and plants. In fact, the regulator role of temperature is commonly recognized in physiology as well as in germination. Indeed, temperature determines the germinability of seeds by determining its rate. It removes primary and secondary dormancy and temperature also induces the second dormancy [11].

Light is primarily responsible for the effect inducing germination after turning soil. As little as one millisecond of exposure to full sunlight can cause many seeds to germinate and produce seedlings. This principle may be utilized to reduce the use of herbicides in weed management programs. Hence, soil plays the role of light filter [11].

7. Climate and bud burst

Break in dormancy in many plants is triggered by temperature [1]. A sufficiently high temperature is indeed needed to make bud bursts. This would occur due the expansion of internodes and leaves formerly formed in previous season. This high temperatures is almost needed for the newly formed buds due to the apical meristem getting activity resumed. On the other hand, the burst of dormant buds is caused by the elongation of the internodes following to cell expansion [12]. The young buds are very sensitive to coldness and may be severely damaged if they burst early in winter. This probable damage of buds have an inevitable repercussion on the crop.

Although, almost factors controlling the phenology of bud burst are poorly understood, bud burst has a particular timing controlled by some climatic and non-climatic factors including:

- **Air temperature**: the temperature of the air affects with its values summation since the mid-winter as well as with its high values in the last part of winter. In fact, there is a correlation between the daily mean temperature and the number of buds bursting per day [12]. However, a need for chilling is indispensable for the bud burst and similarly for seedling.

- **Soil temperature (Root temperature):** there is a positive correlation between bud burst and temperature of the rhizosphere. High temperature (25°C) promotes the early bud burst unlike low temperature of about 12°C. This may explains how our ancestors where tricky as they irrigate plants with warm water to activate the bud burst when it delays. This also implies the importance of site selection and soil management. Indeed, drainage due to stony or calcareous quality of soil boost its spring warmth. In contrary, wet soils like clay soil which maintain cold due to their content of water are recommended to postpone bud burst in cool climates, therefore coincidence of new buds with tough condition is avoided and then chance of ripening is increased [12].

- **Photoperiod**: In a less degree, the photoperiod influences the bud burst. Some plant receptors could detects the spring arrival basing on the increase of it and then incites the bud burst. While long shilling duration advances the bud burst and short shilling postpones it, the long photoperiod may compensate for short chilling duration.

8. Climate and flowering

Blooming or flowering is a featuring phenophase which usually heralds the arrival of spring season. Flowering is controlled by environmental conditions and developmental regulation. The complexity of this regulation is created by an intricate network of signaling pathways [13]. The plant Arabidopsis is a model for the study of flowering mechanism due to the significant number of environmental factors involved in this process for many other species. In addition, the genetic material of this plant is well developed. Many factors influence the flowering such as photoperiod, growth regulators, insolation and sunlight, circadian clock regulation, temperature, and chromatin structure [13].

8.1 Photoperiod role in flowering control

One of the most important factors controlling flowering time in temperate regions is the duration of the daily light period, or photoperiod. Plant genes involved in sensing the photoperiod were identified by the molecular genetic approaches. These genes encode the proteins responsible of the flowering process and its regulation according the environmental conditions. Other genes encodes the components of light signal pathways and components of the circadian clock.

The **Figure 2** illustrates the relations among factors, genes, and processes involved in flowering phenology relating to the photoperiod. The effect of photoperiod on flowering consists of a balance between genes which promote the flowering in green color in the **Figure 2** and in red color those which repress the flowering. All these genes are incited by the sunlight. While Repressive effect is represented by the perpendicular arrows, and overexpression of genes is illustrated but e small upright arrows, the promotion is represented by arrows between genes. Similarly, interaction between proteins is indicated by simple lines. The expression of genes controlled by circadian clock is noted by arrows from it whereas arrows to the clock are for the process of lengthening the period by the gene [13].

8.2 The role of the circadian clock and the central oscillator in flowering

Biological Phenomena are almost systematic and have standard rhythm, timing and cyclicity. These are all controlled and adjusted by the biological clock or the circadian clock. This rhythmicity is at all living beings and in every division of time from one second to one year. The genes controlling the circadian clock have been identified by scientist as to be the gene (per, frq, clock, tau) [14].

The period of circadian clock is almost one day without being affected by transition between day and night, dark and light. This clock is observed in all functions and biological elements from stomat, CO2 assimilation, and gene transcription to the clearly observable phenophases like defoliation, bud burst and flowering. This rhythmicity and circadian are believed to be created by the cyclicity of environmental phenomena and hence underwent a selection pressure. Indeed this circadian clock serves the harmony between environment with it climates and the phenology and therefore permit to living beings to efficiently anticipate their environment changes in particular periodic variations or more appropriately the

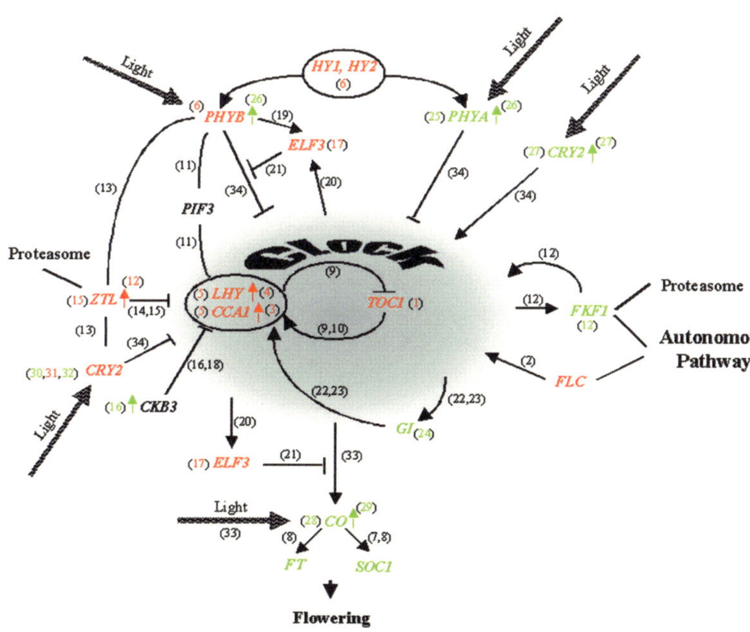

Figure 2.
Signaling pathways involved in the regulation of flowering by photoperiod in Arabidopsis ***source:*** *Reference [13].*

seasonality of climatic elements including temperature, insolation by its intensity and photoperiod, humidity, precipitations [14]. This implies the constant quality of this circadian clock and meanwhile poses a problem of adaption to the unexpected climates changes. Thus, many vulnerable species may have been extinguished due to this effect.

Three outstanding interrelated types of this rhythmicity are characterized in the circadian clock including the input pathways which are relating to the daily cycle of light and dark and adjusting the clock mechanism to it. Secondly, a central oscillator which is responsible to keep the mechanism of 24 hours' time. And output pathways for specific process as the thirds category of this circadian clock [13].

8.3 The role of the vernalization on flowering

As a commonly used technique, vernalization is used since a long history of agriculture. First discovered for plants which were planted in spring, then they needed some cold to germinate and to pass from vegetative life to productive life. Indeed, vernalization is almost related to flowering and fructification in particular. And it is defined as subjection of seeds and seedlings to coldness or chill in order to promote and hasten the growth and the flowering of plants.

In fact, plants are very sensitive to their environment and constantly adapting to the environmental variations by for example dormancy or overwintering during adverse conditions period which is a strategy to withstand them. However, in spite of its toughness, winter cold is in the other hand mandatory and indispensable for plant growth and flowering whence the principle of vernalization. This vernalization is also an adaptive trait to prevent flowering before the spring arrival with its favorable conditions. Genetically, vernalization is merely the inhibition of genes which repress flowering particularly in Arabidopsis and cereals [15].

Exposure to low temperatures for several weeks will often accelerate flowering. Susceptibility to this treatment can differ markedly between varieties of a species. Therefore winter seasons is likely to an inevitable period in the plants life cycle because without long exposure to cold plants do not flower not pass to the reproductive and productive phase of their life. During winter plants are in vegetative growth phase with the minimum of activities [13].

Figure 3 illustrates process, mechanism and all involved factors and genes in the flowering phenomenon, especially promotion of this flowering process is indicted in blue and the red color illustrates the components and genetic interaction which repress flowering including FLC (Flowering Locus C) and its relatives pointed as FLC clade. It inhibits the expression the flowering genes or properly named the floral integrator genes which are FD, FT, and SOC1 (Suppressor Of CONSTANS1). The photoperiod pathway passes by CO (CONSTANS) to induce the FT which is a protein working as a mobile signal of flowering. In the floral meristem, FT in partnership with FD protein activate SOC1 as well as SAP (sepalat), FUL (fruitful), and AP1 (apetatalata1) which are known as the floral meristem identity genes and which induces the floral meristems that will develop into flowers. Accordingly, Flowering locus C (FLC) and photoperiod pathway are antagonist.

In the one hand, Autonomous pathways genes partially determine the expression of FLC. In the other hand, the prolonged winter and cold further repress the FLC remodeling of chromatin. In addition, flowering is promoted by the plants hormones class of gibberellin which activates SOC1 along with the floral meristem-identity gene LEY (LEAFY) [15].

As illustrated by **Figure 4**, vernalization, which repress FLC expression, along with the autonomous pathway genes induce flowering [13].

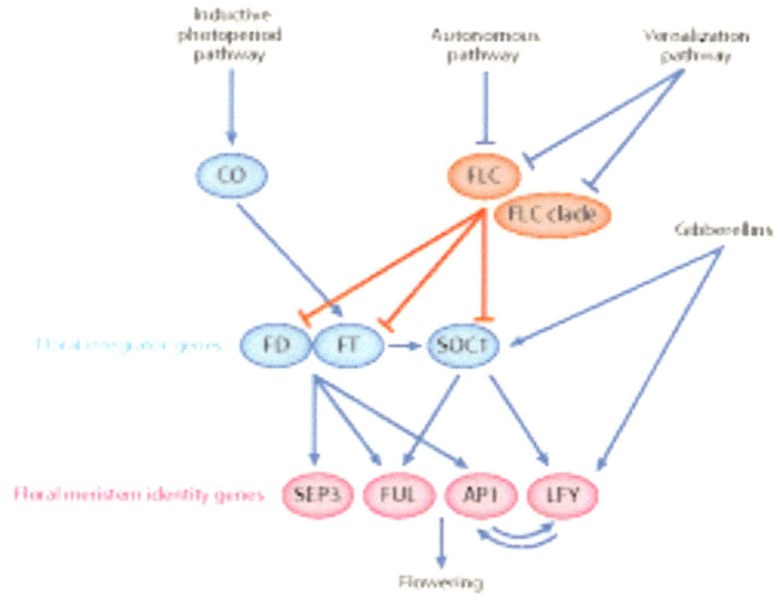

Figure 3.
Outline of flowering pathways in Arabidopsis source: Reference [15].

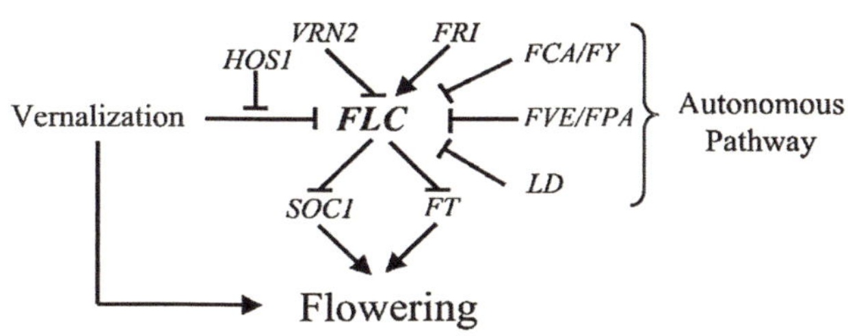

Figure 4.
The effects of Vernalization and the autonomous pathway on flowering time, emphasizing the central role of FLC ***source:*** *Reference [13].*

9. Climate and fructification

The making of a fruit is a developmental process unique to plants. It requires a complex network of interacting genes and signaling pathways which consists of series of reaction launched by an environmental signal such as light, photoperiod, and temperature. Generally, fructification goes through three stages which are first the set of fruit which follows the pollination of flowers. Then the fruit development stage and finally the fruit ripening which received the most attention of researcher of the field. This is due its importance in commercialization an economy since it the quality of fruits is the most attractive feature for the customer. In fact the process of ripening activates a series of biochemical reactions that make the fruit edible and desirable to the consumer [16].

A detailed illustration of this process is provided by the following **Figure 5**. It is merely a scheme of activating the ripening related genes to be expressed into enzymes charged each of which by one or more of the various ripening pathways such odor, color or softening. This whole process is controlled and adjusted by hormonal and environmental signals. It is to be noted that ethylene as a plant hormone plays a major role in this process of ripening and fruit development [16]. Indeed,

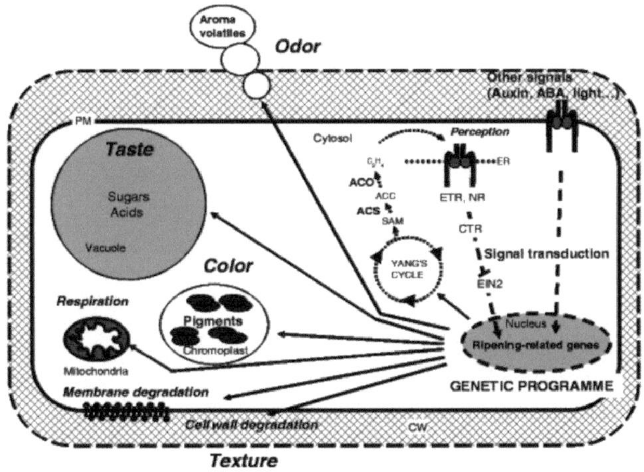

Figure 5.
Schematic representation depicting the molecular mechanisms controlling the ripening of climacteric fruit
source: *Reference [16].*

fruits whose ripening is related to ethylene and a respiration increase are called climacteric fruit, such as tomato, apple, pear, and melon. In the other hand, the non-climacteric are featured by no ethylene associated with the increase and peak of respiration during the ripening phase. It is to note that sales men uses this ethylene to preserve fruits during transportation or deposit for long-period by picking fruits prematurely and when ready for exposition for sale they use ethylene to induce ripening. Similarly, it is used to advance flowering before the adverse periods.

10. Climate change and phenology

The obvious delay of defoliation and advance of blooming along with the appearance of butterfly in autumn and disappearance of many species of plants, insects and animals in many regions unequivocally indicate the climate change and warming, and triggered the scientific research and investigation in this respect [17]. In a context of global climate change being a reality, a continuous process that needs to be taken seriously, and a subject which continues to be a topic of hot debate at global conventions, world summits and international conferences and symposia [1]. Indeed, plant phenology is strongly controlled by climate and has consequently become one of the most reliable bio-indicators of ongoing climate change [18]. Thus, aberrations and anomalies in phenology are repercussions of climatic ones and hence are unequivocal and undeniable evidence of the climate change fact and the fact of their occurring irregularly. This may help to assess and predict ongoing and future significant impacts of climate changes on plants and the whole ecosystems. Effectively, it is largely noticed the disappearance of many vulnerable and hyper-sensible species because of the climatic stress, the unpredicted change of weather, and inability of these species neither to withstand adverse conditions nor to adapt. However if erraticity of weather and climate persist, plants would adapt and a new ecotypes will appear. Indeed, the first response to climate changes is through changes in plant phenololgy and phenophase with their timing and durations and this would have a potential impact on the available resources [1].

Accordingly, long phonological records generate authentic data to study the effect of climate change on phenology and the whole ecosystem, environment and nature future. This may includ parameters such as advance or delay in the appearance of leaf, leaf fall, and timing of opening of flowers, and blooming which can be recorded right at the field site for a long period to form a long time series valid and reliable for the scientific analysis and deductions. For instance, increase in level of carbon dioxide in the atmosphere and consequent global warming may have a profound effect on the flowering time of plants [1].

Accordingly, it was highly recommended to work actively in terms of implementation of the climate change scientific findings in the field of the field of agriculture to both preserve and increase crops. Namely, it was proposed to adjust and reconsider the agricultural calendar according to the recent climate changes to avoid damages related to climate hazards. Indeed, planting dates should be judiciously set in such a way that fragile phenophases (e.g. germination, fruiting) do not come across hard climatic periods [17, 19].

11. Climate change phonological impacts and food security

The climate changes are commonly known as frustrating and alarming when first talked about in the first years of the past century and when they first came into existence among scientific community. All scientist talked about inundation by the sea level rise, drought, desertification, natural resources depletion as climate change aftermaths in a very dramatic way as it is the certain end of life on this planet. In fact, climate changes are constantly with clear and remarkable impacts mainly on living beings and foremost plants due their sessile life style directly exposed to climate influences. Furthermore, this created an intimate relation between plants life and climate or the entire surrounding environments.

Thus, plant's life and physiology is utterly dependent to its environment and mainly the climate.

As the plant is the first source of food for all other living beings, the food security is therefore subject to climate changes impacts. Indeed, since all phenol-phases are interrelated and related to climate seasonality and variation, the fructification as the final one is inevitably impacted by the climate changes and consequently crop quantity and quality. This uncovers and emphasizes the direct impact of climate changes on the food security.

Food is basically from cereal crops whose growth and development are dependent to the day length and growing degree days (GDDs) and they are responsive to climatic factors in specific seasons [20].

The global warming hinders the crop growth and development and mainly causes a shift in phenological development of crops and affects their economic yield [20]. This is due to the fact that the rise of temperature and warm winters are indeed against vernalization which, as previously posited, mandatory for flowering and fructification. This implies that there would be only vegetative life and neither reproduction nor production for plants. Hence, the impact of climate changes is decreasing production in favor of phonological and only vegetative plants.

As climate warming is global and unavoidable phenomenon, the unique solution for food security in this context consists of an adaptive strategy and agronomic management through breeding of climate-adapted genotypes and increasing genetic biodiversity [20]. This is to say that we should make use the field of genetic engineering to develop local species, or more properly and appropriately use more adapted species brought from already temperate regions. This last alternative is more recommended to avoid transgenic organisms and crops whose use as food is unsecure.

12. Conclusion

This chapter is a mandatory background knowledge for everyone interested on nature particularly on plants starting from agricultural professionals to simple farmers, amateurs and herborists. This chapter provides a solid basis for agro-meteorology, and agronomy.

Therefore it is an inevitable part for this very book of agro-meteorology. Actually, the controlling of climate to plant phenophases was emphasized to prove the importance of its knowledge to predict and prevent the probable damages due the climate changes.

For more emphasis, the last two titles were devoted to the impact of climate changes on phenology and consequently on food security.

Acknowledgements

I really owe special acknowledgments and thanks to dear Mrs. Mia Vulovic, the Author Service Manager, for she has encouraged me with insistence, reminded me repeatedly, postponed the deadline for me, and offered me her services with open heart.

Author details

Boubakeur Guesmi
Faculty of Nature and Life Sciences, Univerity Zian Achour of Djelfa, Djelfa, Algeria

*Address all correspondence to: aboubakeurscience@gmail.com

References

[1] Subrahmanyam A. S, Murthy K. S. R: Phenology and climate change. Current Science. 2005; 89(2): 243-244.

[2] Haggerty, Brian P, Mazer S. J. The phenology handbook, a guide to phenological monitoring for students, teachers, families, and nature enthusiasts. 2nd ed. Abridged. University of California, Santa Barbara; 2008. 44p.

[3] Oscar Gordo, Juan José Sanz: Phenology and climate change: a long-term study in a Mediterranean locality. Oecologia. 2005; 146: 484-495. DOI:10.1007/s00442-005-0240-z

[4] Meifang Zhao, Changhui Peng, Wenhua Xiang, Xiangwen Deng, Dalun Tian, Xiaolu Zhou, Guirui Yu, Honglin He, Zhonghui Zhao: Plant Phenological Modeling and its Application in Global Climate Change Research: overview and future challenges. Environ. Rev. 2013; 21: 1 -14.DOI:org/10.1139/er-2012-0036

[5] Michael A. White, Nathaniel Brunsell, Mark D. Schwartz. Vegetation Phenology in Global Change Studies. In: Schwartz (ed.), Phenology: an integrative environmental science, Kluwer Academic Publishers; 2003. p. 453-466. DOI: 10.1007/978-94-007-0632-3_28

[6] Michael Fenner: The phenology of growth and reproduction in plants. Perspectives in Plant Ecology, Evolution and Systematics.1998;1/1:78-91

[7] Wim Soppe, Leonie Bentsink: Dormancy in Plants.in: eLS;2016,p.1-7. DOI: 10.1002/9780470015902.a0002045.pub2

[8] Finch-Savage W.E, Footitt S: Regulation of Seed Dormancy Cycling in Seasonal Field Environments. In : James V. Anderson editor. Advances in plant dormancy. Springer; 2015.p.35-44. DOI: 10.1007/978-3-319-14451-1_2

[9] Gunnar Öquist, Norman P.A. Huner: Photosynthesis of Overwintering Evergreen Plants. Annu. Rev. Plant Biol. 2003; 54:329-55. DOI: 10.1146/annurev.arplant.54.072402.115741

[10] Iqbal N, Masood A, Khan N A: Analyzing the significance of defoliation in growth, photosynthetic compensation and source-sink relations. PHOTOSYNTHETICA. 2012; 50 (2): 161-170. DOI: 10.1007/s11099-012-0029-3

[11] Derek Bewley J., Kent J. Bradford Henk W.M. HILhorst. Hiro Nonogaki. Seeds, Physiology of Development, Germination and dormancy. 3ed. Springer.2013; 407p.

[12] Understanding grapevine growth. Bud dormancy and budburst. 2010; 2p. https://www.awri.com.au/wp-content/uploads/1_phenology_bud_dormancy_and_budburst.pdf

[13] Aidyn Mouradov, Frédéric Cremer, George Coupland: Control of Flowering Time: Interacting Pathways as a Basis for Diversity. The Plant Cell. 2002; 111-130. DOI: https://doi.org/10.1105/tpc.001362.

[14] Björn Lemmer. Circadian Rhythms. In : Encyclopedia Psychopharmacology. 2nd ed. Publisher: Springer-Verlag, Heidelberg Editors: Ian P Stolerman, Lawrence H. Price; 2015. DOI: 10.1007/978-3-642-36172-2_287

[15] Dong-Hwan Kim, Mark R. Doyle,Sibum Sung, Richard M. Amasino: Vernalization:Winter and the Timing of Flowering in Plants. Annu. Rev. Cell Dev. Biol. 2009; 25: 99-277. DOI: 10.1146/annurev.cellbio.042308.113411.

[16] Bouzayen M, Latché A, Nath P, Pech J. C. Mechanism of Fruit

Ripening. In : Plant Developmental Biology - Biotechnological Perspectives; 2010. Vol 1.p.319-339. DOI:10.1007/978-3-642-02301-9_16

[17] Boubakeur Guesmi et al: Analysis of the air temperature records of Djelfa's meteorological station from 1975 to 2014 'the reality of Djelfa's climate warming'. International Journal of Global Warming. 2017; 12(1): 66-84. DOI: 10.1504/IJGW.2017.084015

[18] Oscar Gordo, Juan Jose' Sanz: Impact of climate change on plant phenology in Mediterranean ecosystems. Global Change Biology. 2010; 16 : 1082-1106, DOI: 10.1111/j.1365-2486.2009.02084.x

[19] Boubakeur Guesmi. La steppe algérienne dans le contexte des changements climatiques (cas de Djelfa-Algérie) (EN : the Algeiran steppe in the contexte of climate changes (Case of Djelfa- Algeria)) [Doctorate thesis]. Agronomy department: University of Ouargla; 2016.

[20] Fatima, Z., Ahmed, M., Hussain, M. et al. The fingerprints of climate warming on cereal crops phenology and adaptation options. Sci Rep 10, 18013 (2020). https://doi.org/10.1038/s41598-020-74740-3

Chapter 4

Changes in the Agro-Climatic Conditions in Bulgaria at the End of the 20th and the Beginning of the 21st Century

Valentin Stoyanov Kazandjiev and Veska Anastassova Georgieva

Abstract

Agriculture is one of the most sensitive sectors of the economy. Therefore, climate change has the greatest impact on agricultural production. Despite the achievements of modern agricultural science and the development of agrotechnologies, the importance of meteorological conditions for the size and quality of agricultural products cannot be compensated. Agriculture in Bulgaria is carried out under conditions of limited and insufficient moisture. By the end of the last century, the studied trends of climate change in the area of agricultural production have led to a deterioration of agro-climatic conditions. The temperatures increase and decreasing rainfall and the uneven nature of their distribution cause short-term manifestations of this impact consisting of high variability, including extreme weather events and long-term manifestations consisting of changes in the agro-climatic characteristics of an agricultural area. Since the beginning of the 21st century, we have witnessed annual weather and climate records, both globally and nationally. The frequency and amplitude of extreme weather events is increasing every year. The World Meteorological Organization (WMO) is emphasizing 2019 as the fifth warmest year since the beginning of the 21st century. This paper presents the results of the changes study in agro-climatic conditions in the period 1986-2015 compared to the reference period 1961-1990, as a result of changes in the values of main meteorological elements at the end of the last and the beginning of the present century. The results of the research are necessary for decision-making, both at scientific and management level, in risk assessment and preparation of measures for adaptation to climate change, and directly in agricultural practice - for the choice of crops, varieties and hybrids and in the choice of technological solutions.

Keywords: climate change, deviation, temperatures, rainfalls, vegetative season duration, degree days

1. Introduction

Agriculture, as a branch of our national economy, operates entirely under the open sky. For this reason, climate change and fluctuations and climate anomalies have a very significant impact on the growth, development and productivity of agricultural crops. Climate change affects not only the quantitative but also the qualitative indicators of agricultural production. The development of agricultural practices and the application of precision and organic farming require in-depth knowledge of weather and climate, meteorological and climatic features and agro-climatic resources, in order to effectively and timely manage the processes to achieve maximum results.

The assessment of the influence of various factors shows that the most important for food production are the hydrothermal conditions or more precisely the balance between temperature and humidity conditions. In the agricultural zone of Bulgaria the conditions of humidification are limiting, and they are determined by the balance between temperatures and precipitation. The condition of the main meteorological elements determines the quantitative indicators of the hydrothermal conditions. The established trends of their change toward the end of the last century in Bulgaria showed an increase in the average daily air temperatures, an increase in the minimum temperatures [1–3] and a decrease in precipitation [1, 4].

In order to obtain high yields of high quality agricultural products, the role of soils and their fertility, varieties and hybrids of crops, their biological potential and resistance to extreme weather fluctuations - drought, frost, over wetting and heat waves, which are a prerequisite for stress and reduced productivity. Also, the resistance of cultivated plants to diseases and pests, because when environmental conditions are optimal for plants, they are optimal for the development of weeds, diseases and pests.

The knowledge and combination of the whole complex of biotic, abiotic, economic and technological factors [5, 6] against the background of the changing climate and climatic anomalies with increasing frequency are subject to research by the National Scientific Program - "Healthy foods for strong bio economy and quality of life", and the causal links between agro-climatic and soil conditions and the biological and genetic characteristics of the main types of crops for our conditions in the last 30 years are the subject of research by the project "Agricultural ecosystems adapted to climate change", which is part of the scientific program.

The aim of this study is to discover the changes in agro-climatic resources, as a result of changes in the main meteorological elements, at the end of the last century and the beginning of the present.

2. Experimental data

For the preparation of this up-to-date development, which presents the state of the hydro-thermal conditions in the period 1986-2015 from an agro-meteorological point of view, daily data on the basic meteorological elements were used: minimum, maximum and average daily temperatures and 24-hour precipitation for the indicated period of 56 representative meteorological and agro-meteorological stations for the agricultural zone in the country, **Figure 1**.

The meteorological elements measured also include some agro-meteorological indices that more accurately define the meteorological conditions, namely: the dates of a steady transition of the average daytime air temperature at 5°C and 10°C in spring and autumn; the duration of period with temperatures above 5°C, also known as a potential vegetative season (PVP) and above 10°C - a real vegetative season (RVP); as well as the absolute minimum and maximum temperatures during the coldest and warmest months of the year - January and July, respectively, the absolute minimum temperatures in spring and autumn as an indicator of the occurrence of frost damage to agricultural production, as well as amounts of the active and effective temperatures and amounts of precipitation for the potential and real

DOI: http://dx.doi.org/10.5772/intechopen.93808

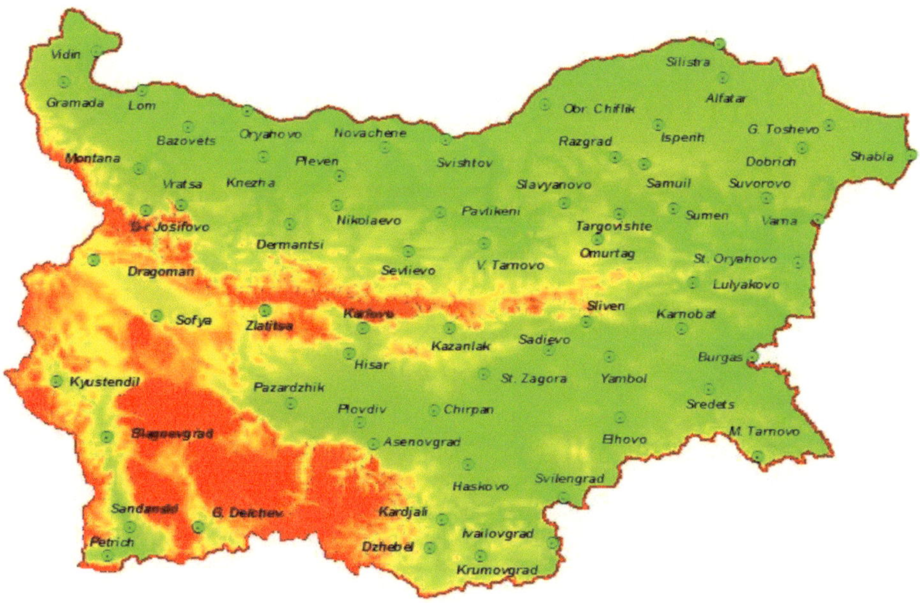

Figure 1.
Map with location of the meteorological and agrometeorological stations in the agricultural zone of Bulgaria.

growing season. In the course of this study, periods of dry spell (10-30 days without rainfall or rainfall less than 5 mm) and persistent drought (more than 30 days without rainfall or rainfall less than 5 mm) were identified. The analysis of the results obtained is a good basis to draw important conclusions, significant conclusions about the agro-climatic resources and agro-meteorological conditions in the NUTS2 administrative territorial planning zones.

3. Results and discussion

The growth, development and productivity of agricultural crops are most significantly influenced by hydrothermal conditions during the growing season. For wintering cereals - wheat, barley and rapeseed - this is the period from October of the previous year to June of the next and for spring and heat-loving from March to September, and for orchards - until October. The analysis of the average monthly long-term values of temperatures during the study period shows that they have increased, in almost all regions and months, compared to the reference period 1961-1990. In precipitation, the trends are not clearly expressed in all months and regions.

3.1 Change in basic meteorological elements during the period 1986-2015 compared to the reference period 1961-1990

3.1.1 Deviation of the average monthly temperature from the norm

The analysis of the deviations of the average multi-annual monthly temperature by months and seasons for the research period (1986-2015) was made by administrative-territorial planning zones, **Figure 2**. In all agricultural areas the temperature deviations from the reference values in December are predominantly negative. Exceptions are some of the stations located in Northwestern and North-Central Bulgaria, where the average monthly temperature for 1986-2015 is higher than the

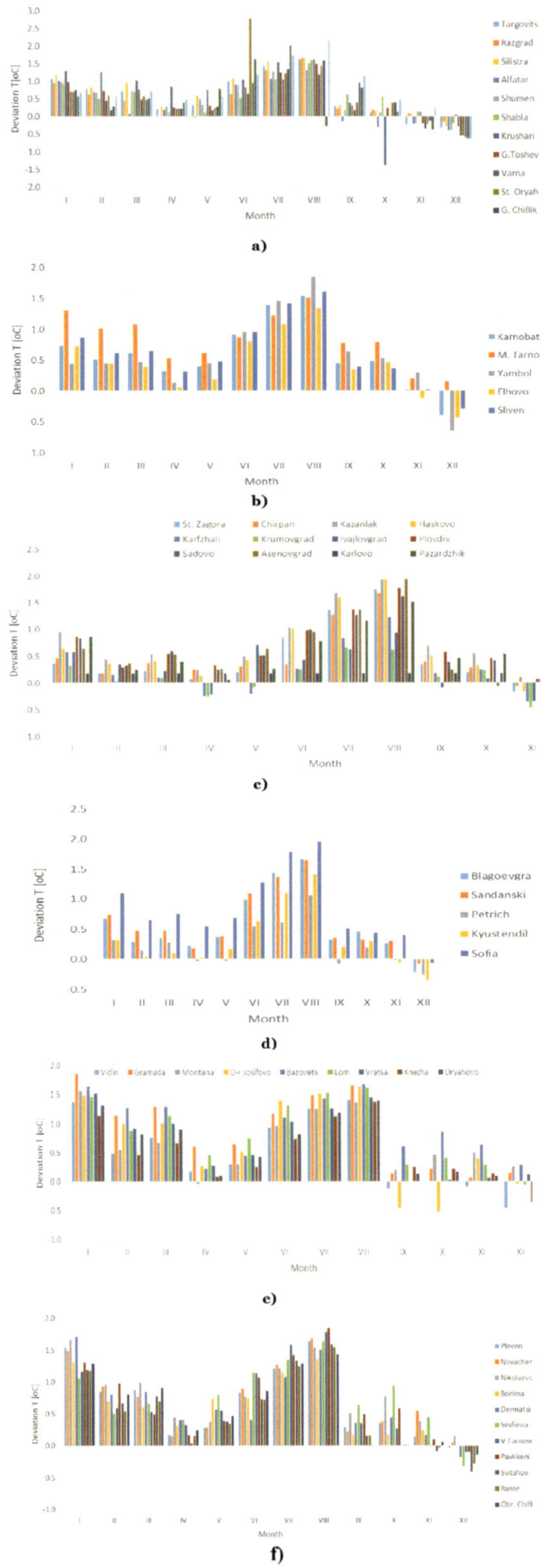

Figure 2.
Average monthly temperatures for the period 1986-2015 and deviations in comparison with 1961-1990 on different regions: (a) Northeastern Bulgaria; (b) Southeast Bulgaria; (c) Central South Bulgaria; (d) Southwestern Bulgaria; (e) Northwest Bulgaria and (f) Central North Bulgaria.

same during the reference period. The most negative temperature deviations are in Northeastern and Southern Bulgaria, reaching -0.6°C.

In January and February the trend is different from that in December. In both months during the study period it was warmer than the reference. In January in Northwestern and North-Central Bulgaria the deviations are higher than 1°C, varying from 1.0°C in Sevlievo and Knezha to 1.9°C in Gramada. In the other regions the deviations vary between 0.3°C in Krumovgrad and 1.3°C in Krushari and M. Tarnovo. In February, the positive trend of deviations is maintained, but the deviations are lower than in January. Without application of the average monthly temperatures in February it remains in Southwestern, Kyustendil and South Central Bulgaria, Krumovgrad, in the other stations the positive deviations vary between -0.1°C, Petrich and 1.3°C, Krushari. The winter in Western Bulgaria during the study period is warmer compared to the winter of the reference period. In Central Bulgaria the trend of positive deviations of the average long-term monthly temperatures is similar to that in Western Bulgaria, but the values of these deviations are lower than in the western regions of the country.

In conclusion, it should be emphasized that despite the large values of the positive deviations, which is evidence of warming, January remains the coldest month of the year in the period 1986-2015. In agro-climatic terms, winters become milder and relatively snowless.

During the spring months of March, April and May (MAM) the deviations of the average monthly temperatures remain positive compared to those for the same months during the reference period in Western, Eastern and North-Central Bulgaria. In South-Central Bulgaria - Kardzhali, Krumovgrad and Ivaylovgrad the deviations of the average monthly temperatures in April and May are negative by 0.2-0.3°C, i.e. in those places it is colder than in the period 1961-1990. In March it is warmer everywhere compared to the reference period, as in Northern Bulgaria the deviations are larger and vary between 0.4°C, Razgrad and 1.3°C in Bazovets. The average deviation in Northwestern Bulgaria is 1.0°C, in North Central 0.7°C, and in North-Eastern 0.6°C. In Southern Bulgaria the largest deviation is in Southeastern Bulgaria, 0.6°C. In South Central and Southwestern Bulgaria it is 0.4°C.

The average monthly air temperature in April has the smallest positive deviations in spring. The averages for the 6 regions vary between 0.1°C in South Central Bulgaria and 0.3°C in East and North Central Bulgaria. In May the average deviations by regions vary between 0.4°C and 0.5°C in Northern Bulgaria and 0.3°C and 0.4°C in Southern Bulgaria.

In conclusion, it can be said that the trend toward warming continues in the spring, but the values of the deviations are smaller than those in January and February.

During the summer months of the study period - June, July, August (Southeast Asia) almost all over the country the deviations of the average monthly air temperatures are positive. The smallest are the average positive deviations in June, between 0.7°C in South-Central Bulgaria and 1.1°C in North-Western Bulgaria. The largest are the positive deviations in August with average deviations by regions, 1.5-1.6°C.

In conclusion, it can be said that during the summer the highest values of positive deviations of the average monthly values of the air temperature compared to the reference period were reported. The largest increase is in August.

The autumn months of September, October and November (SON) during the study period become warmer. An exception is November, when in Northeastern, South Central and Southwestern Bulgaria, the average deviation is negative. The values of these deviations are of the order of 0.1-0.2°C, therefore it can be assumed that these deviations are of the order of the error of this meteorological element.

3.1.2 Deviation of the monthly amount of precipitation from the norm

The size and sign of the deviations of the amount of precipitation during the survey period 1986-2015 compared to 1961-1990 by months, measuring stations and agro-industrial regions of the country are presented in **Figure 3**.

In Western Bulgaria, the average value of the multiannual (1986-2015) monthly amounts of precipitation during the winter months (DJF) has weak positive and negative values of deviations, whose values are minus 10.7 mm in January in Petrich and 13 mm in December in Gramada and would not have a significant impact on the dynamics of autumn-winter moisture accumulation process.

In Central Bulgaria the deviations of the monthly amounts of precipitation during the winter months (DJF) are analogous to those of the western part of the country. The largest negative deviations were registered in January in North-Central Bulgaria and reached 13 mm in Dermantsi. In December, the positive deviations prevail.

In Southeastern Bulgaria everywhere and for all winter months we have a decrease in precipitation. Malko Tarnovo is impressive, where the winter precipitation has decreased by 76 mm for the study period compared to the reference. At the same time, in the northeastern part of the country in December and January the deviations are predominantly positive, while in February they are negative. The positive deviations in December reach 17.5 mm in St. Oryahovo and in January 10.2 mm in Alfatar.

In conclusion, it can be said that in December the precipitation in Northeastern Bulgaria increased on average by more than 10%, and in January in Southeastern Bulgaria it decreased by 18%. The average change in the other areas is insignificant. In February in Western Bulgaria the decrease of the rainfall is between 7 and 8% of the monthly norm, in North Central Bulgaria a decrease of 10%, while in South Central Bulgaria we have a slight increase. The largest decrease is in Eastern Bulgaria, where in the Northeast the decrease is by 13%, and in the Southeast by 11%.

In March, only negative deviations were reported in the stations of Northwestern Bulgaria. In the other areas there is both a decrease and an increase in precipitation. The average deviation by region is positive in the other regions. The decrease in Northwestern Bulgaria is 6% of the March norm. In the other regions the increase is between 2%, North Central and 12%, in Northeastern Bulgaria.

In April, an increase in rainfall was observed in the western regions. In Northwestern Bulgaria change is average of 5% and in Southwestern Bulgaria, by 12% of the climatic norm for April. The largest is in Petrich, where this increase is 30%. A decrease was registered in the other regions of the country. In the central regions it is on average between 4 and 8% of the norm for April. In the South-Central region the decrease is bigger, as in Krumovgrad and Ivaylovgrad, the decrease reaches 12% and 17%, respectively. The most significant is the decrease of the precipitation in South-Eastern Bulgaria, where the average value of the deviations reaches 12% of the monthly norm, and in M. Tarnovo it reaches 22%.

In May there was a clear decrease in the monthly amounts of precipitation by between 6 and 9% for the different regions of the country. An exception is Petrich, where there is a slight increase in the amount of precipitation. The largest decrease is observed in Alfatar 21% (NE) and Nikolaevo 15% (CN).

It can be said that in the spring (MAM) both negative and positive deviations of the monthly precipitation amounts during the studied period compared to the reference period are observed. The average change in the monthly amounts of precipitation by regions reached a 12% increase in March in Northeastern Bulgaria and in April in Southwestern Bulgaria and a decrease in Southwestern Bulgaria in April. By stations, the change increased to 30% of the monthly norm in Petrich in April and a decrease of 21% in Alfatar in May.

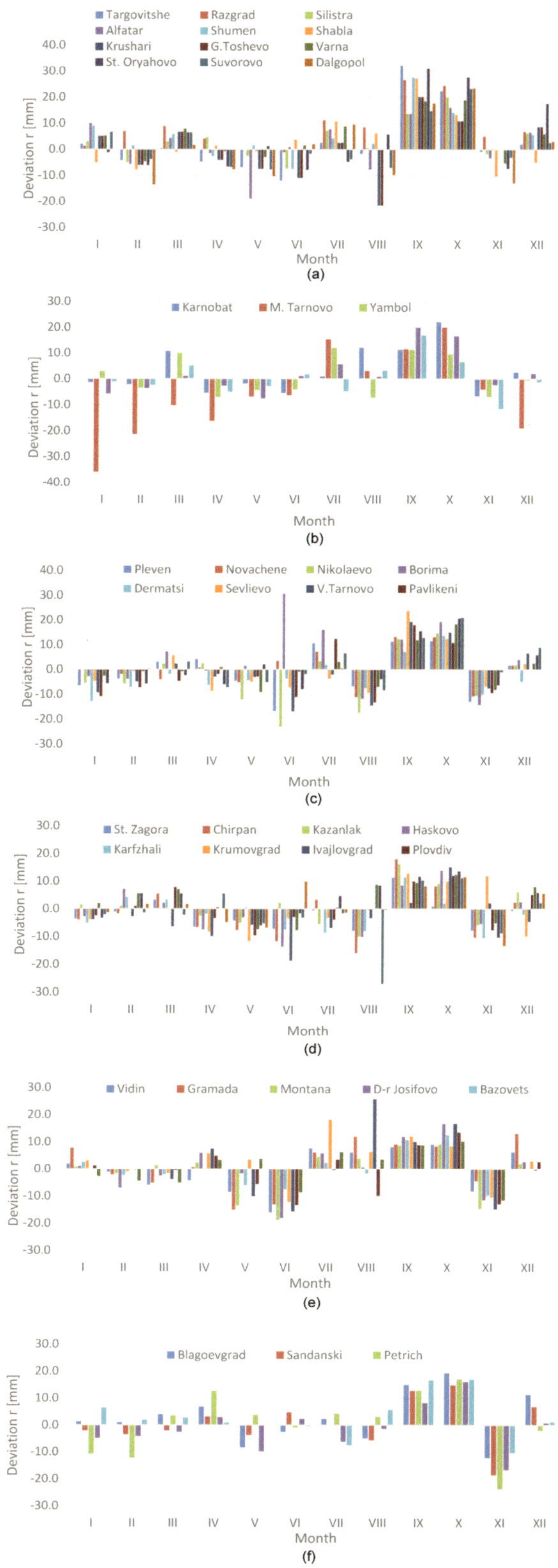

Figure 3.
Deviations in average multiannual monthly rainfalls for the period 1986-2015 compared to the reference period 1961-1990: (a) Northeast Bulgaria; (b) Northwest Bulgaria; (c) Central North Bulgaria; (d) Central South Bulgaria; (e) Southeast Bulgaria and (f) Southwest Bulgaria.

Summer (JJA) is also characterized by various changes in the monthly amounts of precipitation. In June, negative deviations were observed in all regions except Southwestern Bulgaria. They are most significant in Northwestern and North-Central Bulgaria, where the average decrease reaches 17% and 11% of the monthly norm, respectively. At stations in Northwestern Bulgaria the decrease is between 10% in Bazovets and 24% in Vidin. In North-Central Bulgaria the reduction reaches 27% in Nikolaevo. In northeastern Bulgaria the deviations are negative, except for Dalgopol, but the average value is insignificant 5% of the monthly norm. By stations, however, the reduction reaches 18% in Krushari. The decrease of the average monthly amounts of precipitation during the studied period compared to the reference one in South Central Bulgaria is by almost 10%. By stations, in Pazardzhik, Plovdiv, Kardzhali and Kazanlak the change is a slight increase, but in Krumovgrad, Sadovo, Haskovo, Chirpan and Sliven the decrease is between 13% and 21% of the monthly norm. In Southwestern Bulgaria the positive deviation is predominant, as in Sandanski the increase reaches 10% of the monthly norm. In July in Northern and Southeastern Bulgaria there is an increase in precipitation by between 9 and 12% of the monthly norm for the month. In South-Central and South-Western Bulgaria there is a slight decrease. By stations, the increase in average monthly amounts reaches 20% in Razgrad, 24% in Varna, 36% in Shabla and 44% in Lom in northern Bulgaria and 26% in Yambol and 42% in Tarnovo in southwestern Bulgaria. A significant decrease in the monthly amounts of precipitation was reported in Haskovo, 20%; Krumovgrad, 18%; Ivaylovgrad, Sofia and Kyustendil, 12% and Chirpan, 11%.

In August, everywhere, except for Northwestern Bulgaria, a decrease in the monthly amounts of precipitation is reported. It is most significant in North Central Bulgaria, where the average deviation is 19% of the monthly norm and in South Central with a deviation of 14%. In Northwestern Bulgaria there is an increase of 12% of the monthly norm, and in Southeastern Bulgaria by 6%. By stations the biggest decrease in Karlovo, Nikolaevo, and Krushari. The most significant increase in Vratsa, St. Oryahovo, Karnobat, Sadovo, and Asenovgrad.

The most significant change in the monthly precipitation amounts is observed in autumn (SON). In September and October there is an increase in precipitation everywhere, the largest being in Northeastern Bulgaria. There, in September, the increase reached 60% of the monthly value during the reference period, and in October 55%. The average deviation in the other regions fluctuates between 24% in October in South-Central and 45% also in October in South-Central Bulgaria. By stations the largest increase is observed in September in St. Oryahovo, 84%, Targovishte and Shabla 79%.

In November, the deviations are negative everywhere. The decrease in Eastern Bulgaria is insignificant, around and below 10%. But in other areas it varies between 11 and 25% of the monthly norm. In Western Bulgaria the average deviation is 21-25% of the norm, and at stations in Sandanski, 30%; in Petrich, 29%, in Montana 28%, and in Knezha, Pleven and Kyustendil 26% of the monthly norm.

The change in the average monthly precipitation amounts during the survey period 1986-2015 compared to those during the reference period is different during the months of the year. During the off-growing season, there is a predominantly insignificant decrease in the average monthly amounts of precipitation. The more significant decrease in Oryahovo, Dermantsi, Pavlikeni, Shabla, Dalgopol, M. Tarnovo, Kyustendil and Petrich is impressive, and a decrease between 13 and 19% of the norm. An increase of 13% was reported in Razgrad. The decrease in precipitation during the off-growing season is at the expense of the serious decrease in November. In the remaining months the change in precipitation is insignificant.

The average values of the deviations of the average monthly amounts of precipitation during the growing season are insignificant, but positive in all considered regions. In some stations in Eastern and Southwestern Bulgaria the increase exceeds 10% of the norm - Razgrad, Silistra, Shumen, Shabla, Varna, G, Toshevo, St. Oryahovo, Karnobat, Elhovo, Asenovgrad and Petrich. This increase is at the expense of the increase in September and October in these areas.

If analyze the average multiannual amounts of precipitation for the period 1986-2015 for all 56 representative stations for the agricultural zone of the country and their comparison with the same values for the reference period 1961-1990 shows a predominant increase of the annual amount by 20-100 mm as the largest increase of the average multiannual amount was measured in Razgrad. An increase of this indicator by 50-60 mm was registered in the stations Borima, Shumen, Varna and St. Oryahovo. During the study period there are also individual places where the measured average multiannual precipitation amount is less than the same during the reference period. The values of this decrease are 20-70 mm, the largest decrease was found in M. Tarnovo, and in Ivaylovgrad this decrease is 42, and in Kyustendil 45 mm, **Figure 4**.

3.2 Agro-climatic conditions during the period 1986-2015

3.2.1 Dates of a permanent transition of the average daytime temperatures across 5°C and 10°C in spring and autumn and temperature degree days

The steady transition of the average daytime temperature in spring and autumn at 5°C and 10°C determines the duration of the potential vegetation period (PVP) and the real vegetation period (RVP), respectively. For more accurate determination of these dates, 10 representative stations in the agricultural zone of the country were selected: in Northern Bulgaria - Knezha, Pavlikeni, Obraztsov Chiflik and General Toshevo; in southern Bulgaria - Plovdiv, Stara Zagora and Karnobat; and in Western Bulgaria - Sofia and Kyustendil. In the northern part of the country, the average date for a sustainable temperature transition across 5°C at spring is March 17th, and in the fall below 5 C is November 30th. The duration of PVP is 258 days. In Southern Bulgaria, these transitions are March 17th and December 3th respectively, and the duration of PVP is 261 days. In Western Bulgaria, represented by the high hollow fields of the Sofia, Pernik and Kyustendil fields, the transitions across 5°C in the spring on March 18th, and in the autumn on November 24th. The duration of PVP is 252 days.

The average multiannual dates of average daily temperatures transition across 10°C in spring and autumn are similarly determined and shown in **Table 1**.

The data on the average duration of the PVP and RVP during investigation period (1986-2015) and their deviations in comparison with the same values for the reference period (1961-1990) are presented graphically in **Figure 5**.

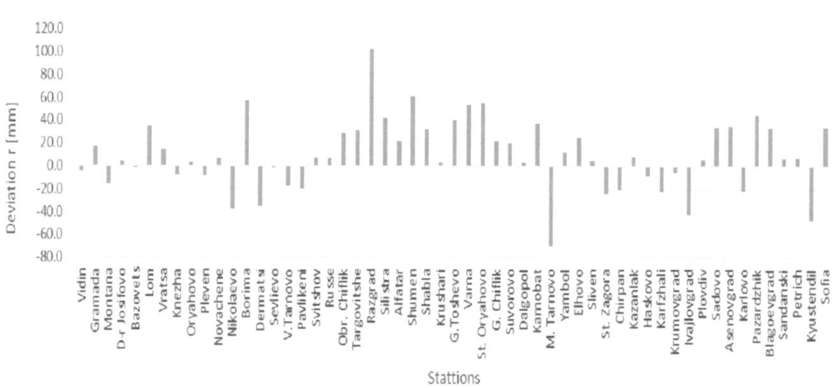

Figure 4.
Deviations between the average multiannual rainfalls for the period 1986-2015 compared to the same values in the reference period 1961-1990 for 55 representative stations located in the agricultural zone of Bulgaria.

Stations	5 °C spr [day]	5 °C spr [date]	5 °C out [day]	5 °C out [date]	Dur 5-5 °C [days]	10°C spr [day]	10 °C spr [date]	10 °C out [day]	10 °C out [date]	Dur 10-10 °C [days]
Knezha	80	21.03	334	30.11	254	105	14.04	305	1.11	201
Pavlikeni	72	13.03	332	28.11	259	99	9.04	311	7.11	212
Obr. Chiflik	74	15.03	333	29.11	259	103	12.04	309	5.11	206
G. Toshevo	78	18.03	338	4.12	261	110	20.04	310	6.11	200
Average for Nord Bulgaria	76	17.03	334	30.11	258	104	13.04	309	5.11	205
Karnobat	81	21.03	334	30.11	253	116	26.04	307	3.11	191
St. Zagora	80	21.03	339	4.12	259	104	14.03	312	5.11	208
Plovdiv	69	10.03	337	3.12	268	91	1.04	298	24.11	207
Average for South Bulgaria	76	17.03	337	3.12	260	104	14.03	306	2.11	202
Kyustendil	75	16.03	328	24.11	253	105	15.04	301	28.10	196
Sofia	79	20.03	328	24.11	252	107	17.04	306	2.11	199
Average for West Bulgaria	77	18.03	328	24.11	252	106	16.04	303	30.10	197
Average for the country	76	17.03	334	30.11	257	105	15.04	307	3.11	202

Table 1.
Average multiannual dates of continuous transition of air temperature over 5°C and 10°C in spring and autumn and duration of PVP and RVP.

The same calculations were made and for sums of active and effective temperatures above 5°C and 10°C are shown in **Figure 6**. Average multiannual sums of rainfall during PVP and RVP during investigation period and deviation in comparison with the referent period are shown in **Figure 7**.

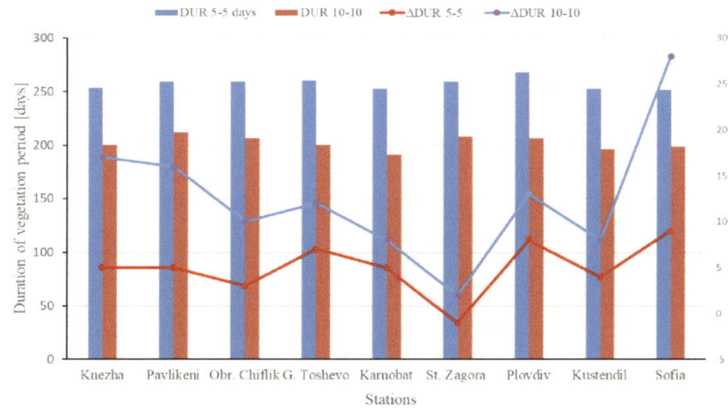

Figure 5.
Duration of vegetative season 5-5°C and 10-10°C for the period 1986-2015 and deviation of duration in comparison with 1961-1990.

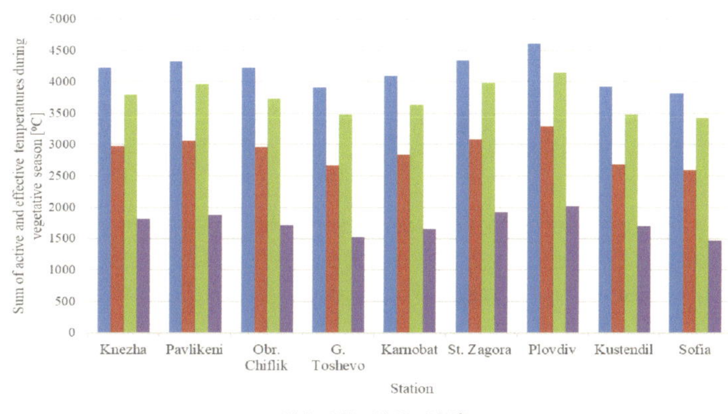

Figure 6.
Average multiannual values of accumulated active and effective temperatures over periods with a sustained transition of temperatures over 5°C and 10°C.

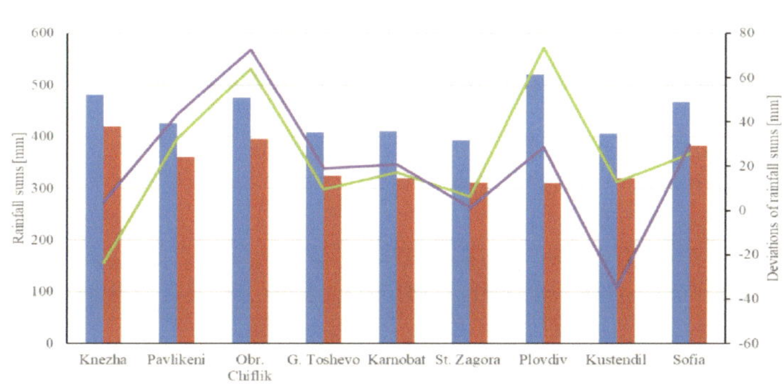

Figure 7.
Average multiannual rainfalls during the potential and real vegetative season in 1986-2015 and their deviation in comparison with the same seasons for the referent period 1961-1990.

4. Conclusions

The presented study is essential for a series of studies within the cited National Science Program and its project - Agricultural Ecosystems - adapted to climate change. The main results for this stage of project development are:

1. The values of the parameters of the climate change for the agricultural zone of the country during the studied period 1986-2015 for 56 meteorological and agrometeorological stations are obtained;

2. The sign and the amount of the deviations in the agricultural areas of the average multi-annual monthly values of the temperatures and the monthly amounts of the precipitation for the period 1986-2015 compared to the reference period 1961-1990 have been determined.

3. The data obtained show larger positive values of the temperature deviations in January, February, July, August and September and negative values of these deviations in December.

4. In case of precipitation - the positive and negative deviations by months are small in value, with the exception of Obraztsov Chiflik and Plovdiv. The non-uniform nature of the distribution of precipitation by seasons increases and dry winters, dry beginning of spring, rainy June and prolonged 70-90 days of summer drought, followed by wet October and November, become more frequent. These features of the climate are a serious challenge for specialists in genetics and selection in creating varieties that can withstand periods of drought;

5. The average long-term values of active and effective temperatures for the periods with $t \geq 5°C$ and $t \geq 10°C$ are obtained, which in most regions of the country are 3500-4000°C respectively and over 1500-2500°C respectively. This is a sure indication that thermal conditions are not a limiting factor and allow the cultivation of second crops, as well as the introduction of tropical crops with a short growing season;

6. The in-depth analysis of the hydrothermal conditions is a prerequisite to recommend, over the next 5 years, a gradual but large-scale increase in investments for the construction of modern irrigation systems. The results of the climate scenarios for the next 20-30 years show the tendency of the observed changes to continue, which will make impossible the efficiency of agricultural production under natural conditions of humidification of soils.

Acknowledgements

This work was supported by the Bulgarian Ministry of Education and Science under the National Research Program - **Healthy Foods for a Strong Bio-Economy and Quality of Life** approved by DCM 577/17.08.2018.

Author details

Valentin Stoyanov Kazandjiev* and Veska Anastassova Georgieva
National Institute of Meteorology and Hydrology, Sofia, Bulgaria

*Address all correspondence to: valentin.kazandjiev@meteo.bg

References

[1] Kazandjiev V., M. Degorski, K. Błażejczyk, V. Georgieva, 2015: Agroclimatic conditions in Bulgaria and agriculture adaptation, Europa XXI, No 29, 2015, Warsaw, Poland

[2] Eitzinger J., S. Thaler, S. Orlandini, P. Nejedlik, V. Kazandjiev, V. Vucetic, T.H. Sivertsen, D.T. Mihailovic, B. Lalic, E. Tsiros, N.R. Dalezios, A. Susnik, Christian K.C. Kersebaum, N.M. Holden, R. Matthews 2008. Agroclimatic Indices and Simulation Models, in "Survey of Agrometeorological Practices and Applications in Europe, Regarding Climate Change Impacts", ESF, COST 734, pp. 15-114

[3] Kazandjiev V., 2011. Climate change, Agroclimatic Resources and Zoning of Agriculture in Bulgaria, Journal of Balkan Ecology v.14/4 pp. 365-382. ISSN: 1311-0527

[4] Slavov N., V. Kazandzhiev 2003. Study on the agro-climatic indicators for the onset of spring in Bulgaria, Collection of scientific reports 75 Years of the Forest Institute, pp. 55-60. (in Bulgarian)

[5] Kazandjiev V. 2017. Climate Change – Fundamentals; Agroclimatic Conditions in Bulgaria and Resilience Agriculture Trough Adaptation, in "Implications of Climate Change and Disasters on Military Activities: Building Resiliency and Mitigating Vulnerability in the Balkan Region, NATO SfPS series C: Environmental Security, 158 p.

[6] Hershkovich E. L Agroclimatic Resources of Bulgaria, Publisher, BAS 1984, 115 pages. (in Bulgarian)

Chapter 5

Atmospheric Pollution Interventions in the Environment: Effects on Biotic and Abiotic Factors, Their Monitoring and Control

Nukshab Zeeshan, Nabila, Ghulam Murtaza, Zia Ur Rahman Farooqi, Khurram Naveed and Muhammad Usman Farid

Abstract

Atmosphere is polluted for all living, non-living entities. Concentrations of atmospheric pollutants like $PM_{2.5}$, PM_{10}, CO, CO_2, NO, NO_2, and volatile organic compounds (VOC) are increasing abruptly due to anthropogenic activities (fossil fuels combustion, industrial activities, and power generation etc.). These pollutants are causing soil (microbial diversity disturbance, soil structure), plants (germination, growth, and biochemistry), and human health (asthma, liver, and lungs disorders to cancers) interventions. All the effects of these pollutants on soil, plants, animals, and microbes needed to be discussed briefly. Different strategies and technologies (HOPES, IOT, TEMPO and TNGAPMS) are used in the world to reduce the pollutant emission at source or when in the atmosphere and also discussed here. All gaseous emissions control mechanisms for major exhaust gases from toxic to less toxic form or environmental friendly form are major concern. Heavy metals present in dust and volatile organic compounds are converted into less toxic forms and their techniques are discussed briefly.

Keywords: air pollution, abiotic, biotic, control, devices, sensors

1. Introduction

Environmental pollution (EP) is successfully deteriorating our surroundings. Experts and major stake holders are trying to overcome it. EP not only effecting air, water, land but also plants, microbes, and humans. Atmospheric pollution (AP) is worse because it directly inhales by living organisms, particularly humans. Manmade activities increase the level of air pollution from preindustrial age [1].

Air pollutants are of many types and are classified on their suspension in the environment. Major air pollutants (AP) include suspended particulate which includes dust, fumes, mists, gases, vapors etc. The sources of these pollutants are diesel exhaust, coal fly-ash, mineral dust (asbestos, coal, cement and lime), metal dust (Cu, Fe, Pb and Zn), fumes, acid vapors (H_2SO_4), paints, pigments, black carbon and smoke from oil.

IntechOpen

Ozone level of various sites of Northern Hemisphere has been increased from 10 to 50 ppbv since 1860 [2]. Sulfate aerosols increase from 3 to 4 folds [3].

Pollutants which are suspended in the air are responsible for different diseases like respiratory and Cancer, corrosion of metals and damage of plant biochemistry. Most of these pollutants deposited on the surface of the plants and cause naissance and disturb sunlight interaction with chlorophyll, the scattering of light from these pollutants produce smog and many surface chemical reactions. Many air pollutants are in gaseous form, like oxides of sulfers (SO_2 and SO_3), carbon (CO_2 and CO), and nitrogen (NO_2 and NO_3). Most of organic compounds are also present in the air like Hydrocarbons, Volatile organic compounds, poly aromatic hydrocarbons and different halogens and their derivatives. Chemical, Thermal and photochemical reactions of above mentioned pollutant caused secondary pollutants. A common example of thermal pollution is when the oxidation of SO_2 occurs to SO_3 by thermal reaction. If SO_3 further catalyzed in the presence of Mn and Fe in water than it give rise to sulfuric acid mist. Nitrogen oxides and reactive hydrocarbons when react. They produced ozone, per-oxy-acetyle nitrate (PAN). Some order causing agents are also produced which are known as hydrogen sulfide, carbon disulfide and mercaptan while others are very difficult to define chemically.

Chemistry of atmosphere totally depends upon the chemistry of pollutant present. The activities such as stream of traffic, industrial emissions, cleaning and washing of roads, painting, repairing are the causes of air pollutant generation. Because of their harmful effects, the air pollutants are now major concern of human talks [4]. Noise pollution is also the part of air pollution. Increase of traffic and other anthropogenic activities the noise pollution is increasing day by day. It is also causing swear health effects in humans (high blood pressure, sleeplessness, nausea, depressions are common). The study revealed that air pollution is responsible for

Some important terms and their definitions related to air pollution.

Term	Definition	Citation
Air pollution	Occurrence of dangerous particles and chemical species into the surrounding air beyond the permissible limit is known as air pollution e.g. $PM_{2.5}$, NO_X, SO_X etc.	[6]
$PM_{2.5}$	The inhalable particles present in the air of size 2.5 micrometer or smaller are categorize as PM2.5.	[7]
PM_{10}	The inhalable particles present in the air of size 10 micrometer or smaller are categorize as PM2.5.	[8]
CO	It is a colorless, odorless dangerous gas present in the environment. It is a silent killer and mostly produced in low oxygen.	[9]
CO_2	It is atmospheric gas essential for globe temperature balance but harmful at its high concentration. It is also colorless and 60% denser than air.	[10]
NO	It is known as oxides of nitrogen and known to general public as laughing gas.	[11]
NO_2	Oxides of nitrogen which enter into the environment from burning of fossil fuels.	[12]
SO_2	Atmospheric gas which has pungent smell. It natural sources are volcanic eruption and anthropogenic are fossil fuel exhaust.	[13]
VOC	Compounds of carbon and hydrogen which convert into vapor or gases phase and contaminate the surroundings are known as volatile organic compounds.	[14]
Aliphatic Hydrocarbons	Compounds with sigma bonds and delocalized pi electrons between carbon atoms forming a circle.	[15]

430,000 premature deaths in Europe and almost 10,000 deaths due to noise pollution have been recorded. Due to these statistics air and noise pollution is listed in the top two stressors in the environmental burden disease [5].

According to the above discussion it is clear that the study of air pollution monitoring, their impacts on biotic components and their control strategies are very necessary to discuss in the nut shell.

2. Effects of atmospheric pollutants on biotic components of environment

Figure 1 is showing the impacts of air pollution on Animals, Plants, Humans and Microbes.

The effect of these pollutants on the plants, humans, animals and microbes are comprehensively discussed in the following **Table 1**.

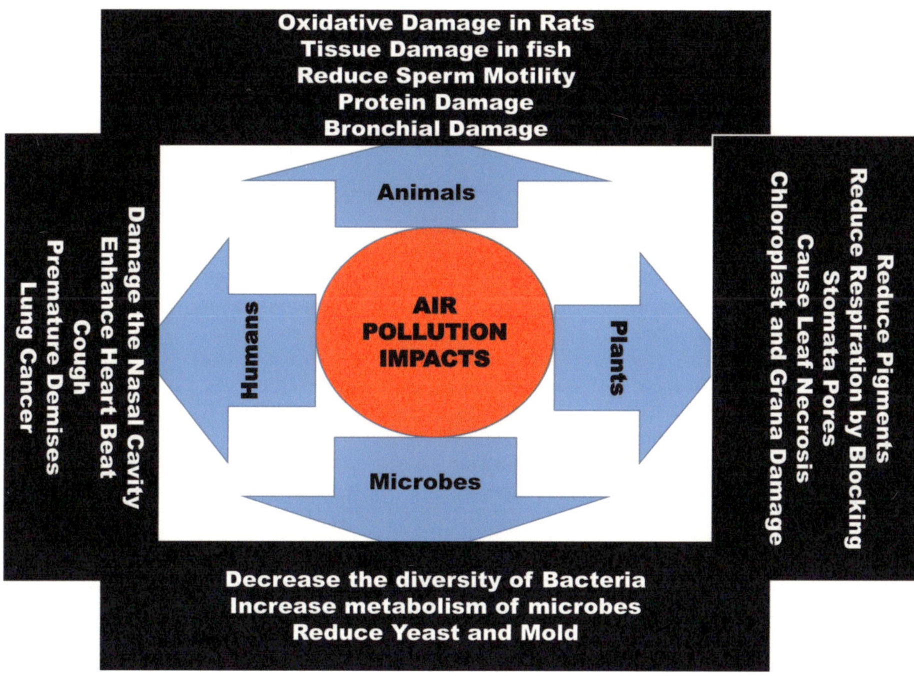

Figure 1.
Impact of air pollution on biotic components of the environment.

3. Air pollution and its control measures

Aerosols are classified into solid [SPM, Dust ($PM_{2.5}$ and PM_{10})], liquid (fumes, mist, vapors) and gaseous (smokes, gases) particles. Air pollutants are categorize the matter which is suspended in the air like road dusts, fumes of chemicals, mists, smoke from different emissions), gaseous pollutants (gases and vapors) and odors producing reagents.

4. Atmospheric pollution monitoring devices

In this **Table 2** the major monitoring devices with their technology is listed for detail review.

Pollutants	Plants	Humans	Animals	Microbes
Dust	Reduce the pigments, enzymatic activity and respiration. Also block the pores in plant leaf [16–18].	Enhance metals intake [19], damage the lining of nasal cavity and enhance secretions [20].	Damage the DNA of animals and increase protein oxidative damage in obese rats [21].	Decrease the diversity of bacteria at highly polluted dust [22].
CO2	C4 plants are less beneficial than C3 plants under high CO_2 environment [23]	Increase sleeping, blood circulation, heart beat when exposed to high CO_2 [24, 25]	High concentration leads to Zn accumulation and tissue damage in fish [26].	Elevated CO_2 effect metabolism, cell structure and diversity of microbes [27].
SO2	Leaf necrosis. Dwarfing and necrosis. 0.05 to 0.5 ppm of SO_2 damage the Spanish and cucumber while apple, barley and wheat are most sensitive to SO_2 [28].	Small exposures causes cough while long exposures causes asthma [29, 30]	Reduce the sperm motility in rats alter the seminiferous tubules in testis [31] and higher exposure induces cardiovascular problems [32]	In the open air fumigation study minute quantity meaningfully reduced respiration in both pine and deciduous litter [33].
NOx	It delays photosynthesis. Concentration of 100 ppm cause spotting to leaf and break down. Oxidative stress increased due to boost in deduced oxygen species [34]	Premature demises cause in humans [35], Long-term contact to NO_x is projected to lay foundation of 2119 respiratory demises and 991 lung cancer demises [36].	High level of protein damage was observed in tree sparrow when exposed to higher level of NO [37].	Not Yet Studied, a strong research gap exists here.
Ozone	Ozone causes foliar injury, reduce the stomatal conductance, enhance the foliar injury and ultimately decrease the plant total biomass [38]	Premature death is the major concern [39].	Ozone exposure to mice evidence the enhance air passage way inflammatory cell infiltration and bronchial hyper-responsiveness as compared to control [40]	Significantly reduced the mold or yeasts in yoghurts ozonated for 60s. *Escherichia coli* O157:H7 count reduced during vacuum cooling droplets of high ozone demand [41].
Aromatic Hydrocarbons	Cause fragmentation of nucleus and mitochondria and mitochondrion, chloroplasts and grana collapse [42].	White matter of left hemisphere reduced and related with slower information. Rapidity during intelligence testing [43].	Consuming the contaminated see food with heavy metals, PAHs and TPHs which is potentially poisonous [44].	Microbes are used mostly to remediate the site contaminated with PAHs.

Table 1.
Effect of the pollutants.

Sr. No	Device	Technology used with pollutant	Citation
1	Portable Monitoring Device for indoor air Pollution	For humidity and temperature complementary metal-oxide semiconductor (CMOS) technology, particulate matter by Laser-based light scattering, volatile organic compounds by Metal oxide gas sensor, CO_2 by Non-dispersive infrared (NDIR), CO by Amperometric gas sensor, light by Infrared-responding photodiode and sound by Electret microphone with amplifier.	[45]
2	Home Pollution Embedded System (HOPES)	Internet of thing (IOT) device which is the grouping of gas semiconductor devices and an Infrared particulate matter sensor.	[46]
3	IoT Based system of Solar Power Environmental Air Pollution and Water Quality Monitoring System	Cheap system for sensing of alcohol, benzene, CO_2, and NH_3. When it is connected it to Arduino then it is able to sense the gases, and provide readings in PPM (parts per million).	[47]
4	A raspberry Pi controlled cloud based air and sound pollution monitoring system with temperature and humidity sensing	It is based on four modules which are 1. Module for monitoring Air Quality Index Monitoring, 2. Module for detection of Sound, 3. Module for Cloud-based Monitoring 4. The Anomaly Notification Module	[48]
5	The Next Generation Air Pollution Monitoring System (TNGAPMS)	Static Sensor Network (SSN), Community Sensor Network (CSN) and Vehicle Sensor Network (VSN) based on the carriers of the sensors.	[49]
6	The Ozone Monitoring Instrument (OMI)	OMI is an ultraviolet/visible (UV/VIS) nadir solar backscatter spectrometer, used to measure UV irradiance, trace gases of tropo-spheric and strato-spheric chemistry.	[50]
7	Wireless distributed sensor networks	It is based on three metal oxides (MO) chemo-resistive sensors for O_3, NO_2 and TVOC, an optical (IR based) total (TSP) sensor, noise sensor and a dual semiconductor sensor for temperature and humidity (RH) measurement.	[51]
8	TEMPO	It measures the spectra required to recover O_3, NO_2, SO_2, water vapors, ultraviolet radiation, and foliage properties.	[52]
9	Amperometric electrochemical gas sensors	It is used for the monitoring of inorganic gases	[53]

Table 2.
Air pollutants monitoring devices along with technologies.

5. Particulate pollution control devices

The burning of diesel causes emissions. These emissions contain toxic gases and particulate matter (PM). Due to which there is a need to control these gases and particulate. For particulate control the diesel particulate filter is used to bind the PM which mostly is the combination of soot particles and organic fraction (soluble). The one bad thing with this system is the accumulation of soot particles in the filter lowers the activity of filtration [54]. To control the particulate matter from commercial cooking three technologies are used. The technologies named as Control technologies (CT) 1, 2 and 3. CT2 is the removal of grease technology which is based on the boundary layer momentum theory. Particulate matter was the significant higher in base line (CT1) than CT2. CT3 technology is Electrostatic precipitator based and is use full to reduce the volatile organic compounds like acetaldehyde and formaldehyde produced during commercial cooking [55]. The efficiency of Electrostatic precipitators is reduced if the temperature of the flue gases increases. The low-low temperature EP (LLTESPs) is more effective in particulate matter removal in coal fired plants. This temperature can be control by using Wet flue gas desulfurization (WFGD) in ESP. The study was conducted to check the effectiveness of the LLTESPs and WFGD. The outlet samples indicate that the concentration of PM decreased with the decrease of temperature. The concentration of soluble ions like mainly SO_4^{-2}, Cl^- and NH_4^+ decreases in the outlet of LLTESPs (0.3 to 0.8 mg/m^3) with respect to WFGD because the addition of gypsum slurry in WFGD (4.7 to 0.8 mg/m^3) [56]. Preventive measures have been taken by individuals to get rid of polluted air. The facemasks are most commonly used by the Chinese people during the extremely high days. The model showed that 100-point increase in air quality index increases 54.5% consumption of facemasks. 187 million dollars could be save if control on air pollution has been achieved and it can be used for the social welfare of the habitants [57]. To combat with the particulate pollution there is a need of the hour to control the emission sources of particulate pollution with improved technologies [58].

6. Gaseous pollution control

CO: Catalytic converter (CC) is a device which is used to convert the hazardous exhaust gases to non-hazardous exhaust gases by using the simple technique of redox reaction. The working of CC is totally based on the catalyst used. For CO control the Silver is used as a catalyst. The more the catalyst, the more the active site and fast reaction.

Catalysts were identified by BET, FTIR, SEM- EDX, XRD, XPS technologies [59]. A study revealed that ZnO–CuO created hetero-composites show selective CO detecting with T_{100} is in close vicinity to T_{opt} to yield simultaneous CO detecting together with its 100% catalytic oxidation for detection devices. The initiated oxygen reacts immediately with adsorbed CO to provide desired CO detecting together with 100% CO oxidation [60]. Evidences showed that oxidative desorption of CO enhance if oxygen species are present. Fast slaking of platinum in water boost the oxidation by two processes. One of the processes is chemical oxidation by using molecular oxygen and other is Langmunir-Hinshel wood surface oxidation [61].

SO2: The usage of segmented multistage ammonia-based liquid spray with different oxidation potentials to remove sulfur compounds from gas. MnO_2 filter are used to absorb the SO_2 from the exhaust. There are many sources of SO_2 production like agricultural heavy machinery, vehicles etc. In this technology MnO_2 along with ozone gas is introduced and found that 90% SO_2 absorption is possible with the addition of ozone. This system also has the ability to improve the NO_2 exhaust [62]. The alternative methods to reduce sulfur emissions are the use of low sulfur fuel, scrubbers to lower the emissions from sulfur rich fuel. Low sulfur in fuel and use of CNG reduce the SO_2 emission [63].

Oxides of sulfur (SO_x) are produced and exhausted during the operations of petrochemical industry and cause harmful effects on environment. One of the technique is sulfur recovery unit (SRU) which is made up of Claus process for removal of huge amount of sulfur removal and afterward a tail gas treatment unit (TGTU) for the remaining H_2S removal (SCOT process, Beavon sulfur removal (BSR) process, and Wellman-Lord process) and flue-gas desulfurization (FGD) processes (once-through or regenerable) [64]. Conversion of H_2SO_4 from SO2, which could be a great impact on reducing pollution [65]. Various approaches for controlling SO_2 emissions include.

7. Wet scrubber

In this technique, SO_2 is absorb by the slurry of an alkaline chemical reagent, and $SO_{2\,(g)}$ is either converted to liquid or solid.

Lime/limestone scrubbing: There are many sorbents but limestone is efficient for desulfurization process. Gypsum scaling is common when the $CaSO_4$ is more than 15%. To avoid this scaling lime stone forced oxidation process is used. In this scaling oxidation of $CaSO_3$ $CaSO_4$ by blustering in the air (usually in the reaction chamber) [66].

Sodium (hydroxide) scrubbing: Sodium scrubbing liquor is very efficient in absorbing emitted SO_2. It is usually used in industrial broilers.

Ammonia scrubbing: It is a unique and new technology, commonly used for desulfurization (DS) of flue gases, in this ammonia is used for DS and commercial grade crop fertilizer is produced. It is currently using by Dakota Gasification Company's (DGC) Synfuels Plant [67]. Further, electrostatic and electro-fabric precipitators are used to remove SO_3 from the flue gases of coal power plants [68].

8. Chlorine emission control

Chlorine emission control technologies are necessary to meet the low emission standards of the USEPA. A study showed that flue gases were samples and analyzed by different emission control technologies (Selective non catalytic reduction, Electrostatic precipitators and fabric filters) and found that 86.1% of chlorine is exhausted in the form of gas. HCl is found significant in samples. The exclusion efficiencies of total chloride are 15.6% by ESP and 19.0–19.7% by FFs, respectively [69].

9. NO_X control

An exhaust system is designed (patent) which has the ability to store NOx at temperature below 200° C and release the NOx above 200° C [70]. Rising trends of Nitrate aerosols were observed in china. The main cause is day time nitrate emissions. These can be controlled if the day time emissions of NH3 and O3 be under-control [71]. NOx emissions are very common from the burning of dried sewage sludge. It is studied that if the combustor physical and operation condition maintained than NOx emissions can be controlled about 75%. Further argued that moderate or intense low oxygen dilution is best suited option to reduce NOx with the cyclone type furnace [72]. Another study suggested that air staging can lead to higher reduction of NOx [73]. NOx can further be controlled from the diesel exhaust by controlling the temperature. It could be 90% less emission if the temperature is minimized. Flue gas treatment with ozone oxidation technology is used to remove NOx. Increase in solubility and bond breakage is the key to success for this technology [74]. The three leading stack gas treatment techniques for NO_x control are catalytic reduction with ammonia, non-catalytic reduction with ammonia, and direct scrubbing of NO with simultaneous absorption of SO_2. The wet processes are much less developed than the dry processes [75].

10. Control of heavy metals pollution in exhaust gases

Study showed that heavy metals show different fate. The control devices which are used in incinerators and other pollution control devices. Some heavy metals like cadmium and plumbum stick in fabric filter ash while chromium, copper and nickel were predominant in the ash present in bottom of the boiler. Zn was found at the bottom and in the ash of fabric filter with a ratio of 07: 03. Though, very minute Hg was found furnace ash, boiler, and SDR and fabric filter; most of Hg crossed through the fabric filter and occurred in an oxidized form. The wet scrubber showed high level control efficacy for mercury which is oxidized, and the addition of commercial stimulated carbon at a rate of 0.2 g/Sm3 resulted in 93.2% mercury removal efficiency [76]. One study revealed that, a high-gravity method using alkaline wastes, i.e., fly ash from petroleum coke, was planned for control of air pollution, containing NO_X, CO_2 and aerosols. Further reacted fly ash can be used for additional cementations material [77].

11. Odor pollution and its control

Odor pollution control is very important for industries and domestic processes because it also caused disputes among neighbors. There are many order producing compounds which includes, organic ammonia, mercaptans and sulfides. Organic and inorganic amines are also very common [78]. NH3 scrubbers are used. The modified scrubber contains two parts. One part use water to remove dust pollution and other part contain dilute acid solution for removal of ammonia and VOCs. Different acidic salts which include aluminum sulfate (alum), sodium bisulfate, potassium bisulfate, ferric chloride and ferric sulfate were found to work as well as strong acids (hydrochloric, phosphoric and sulfuric) for capturing NH3. This technique could result in the capture of a significant amount of the N lost. It also improves the environmental acceptance by the neighbors due to odor control [79].

12. Biodegradation

The degradation of organic pollutants by using the natural force (microorganisms) to water and carbon-di-oxide is known as biodegradation (BD). In artificial technique heat is used but in BD microorganism were utilized. BD efficiently occur at optimum moisture conditions, If plenty of moisture is available than bacteria grow efficiently and BD process speedup and vice versa [80].

13. Conclusions

Anthropogenic accelerated atmospheric pollution is very much dangerous to biotic as well as abiotic factors of the environment. There are different air pollutants ($PM_{2.5}$ and PM_{10}, dust, NOx, Sox, CO, CO_2, and VOCs) have different ways to cause damage to soil, plants, humans, and animals. Sometime this is even lethal for living things and cause pulmonary disorders to even cancers. As pollution is originated from all the anthropogenic activities like industrial processes, power generation and traffic vehicles and are part of economic externalities. These activities cannot be stopped but their life so there are many control technologies which minimize pollutants release into the atmosphere and save the biotic and abiotic components from damage.

Conflict of interest

This not a commercial activity so there is no conflicts of interests between the authors.

Notes/thanks/other declarations

Thanks to IntechOpen for providing opportunity to publish a Book Chapter free of cost.

Appendices and nomenclature

There is no appendices or nomenclature.

Author details

Nukshab Zeeshan[1*], Nabila[2], Ghulam Murtaza[1], Zia Ur Rahman Farooqi[1], Khurram Naveed[1] and Muhammad Usman Farid[3]

1 Institute of Soil and Environmental Sciences, University of Agriculture, Faisalabad, Pakistan

2 Department of Rural Sociology, University of Agriculture, Faisalabad, Pakistan

3 Department of Structures and Environmental Engineering, University of Agriculture, Faisalabad Pakistan

*Address all correspondence to: nukshabz@gmail.com

References

[1] IPCC, C. C., The Scientific Basis. Contribution of Working Group I to the Third Assessment Report of the Intergovernmental Panel on Climate Change, edited by: Houghton, J. T., Ding, Y., Griggs, D. J., Noguer, M., van der Linden, P. J., Dai, X., Maskell, K., and Johnson, C. A., Cambridge University Press, Cambridge, United Kingdom, 2001. p. 1000.

[2] Gros, V.: Background ozone and long distance transport of nitrogen oxides, global change magazine for schools, published by ACCENT, 2006.

[3] Lamarque, J.-F., et al., The Atmospheric Chemistry and Climate Model Intercomparison Project (ACCMIP): overview and description of models, simulations and climate diagnostics, Geosci. Model Dev. Discuss., 2012.(5), p. 2445-2502.

[4] Naveed, K., et al., Air Contamination and Its Impact on Plants, Humans and Water of Pakistan-A Review. J. Appl. Environ. Biol. Sci, 2016. 6(8): p. 32-39.

[5] Khan, J., et al., Road traffic air and noise pollution exposure assessment – a review of tools and techniques. Science of The Total Environment. 2018. (634) p. 661-676.

[6] Meetham, A.R., D. Bottom, and S. Cayton, *Atmospheric pollution: its history, origins and prevention*. 2016: Elsevier.

[7] Zhang, Y.-L. and F. Cao, *Fine particulate matter (PM 2.5) in China at a city level.* Scientific reports, 2015. **5**: p. 14884.

[8] Kim, K.-H., E. Kabir, and S. Kabir, *A review on the human health impact of airborne particulate matter.* Environment international, 2015. **74**: p. 136-143.

[9] from Carbon, N.I. and M.P.J.H. Pain, *Neurological Impacts from Carbon Monoxide Poisoning.* Journal of Headache & Pain, 2016. **1**(3): p. 26.

[10] Liu, Q., et al., *Using carbon dioxide as a building block in organic synthesis.* Nature communications, 2015. **6**: p. 5933.

[11] Larson, C.P. and R.A. Jaffe, *Nitrous Oxide: Yea or Nay*, in *Practical Anesthetic Management*. 2017, Springer. p. 69-79.

[12] Garcia, E., et al. *Reduced Asthma Incidence in Children following Decreased Nitrogen Dioxide Levels in Southern California*. in *AMERICAN JOURNAL OF RESPIRATORY AND CRITICAL CARE MEDICINE*. 2018. AMER THORACIC SOC 25 BROADWAY, 18 FL, NEW YORK, NY 10004 USA.

[13] Schroeter, L.C., *Sulfur dioxide: applications in foods, beverages, and pharmaceuticals*. 2015: Elsevier.

[14] Kamal, M.S., S.A. Razzak, and M.M. Hossain, *Catalytic oxidation of volatile organic compounds (VOCs)–A review.* Atmospheric Environment, 2016. **140**: p. 117-134.

[15] Ghosal, D., et al., *Current state of knowledge in microbial degradation of polycyclic aromatic hydrocarbons (PAHs): a review.* Frontiers in microbiology, 2016. **7**: p. 1369.

[16] Prusty, B.A.K., P.C. Mishra, and P.A. Azeez, *Dust accumulation and leaf pigment content in vegetation near the national highway at Sambalpur, Orissa, India.* Ecotoxicology and Environmental Safety, 2005. **60**(2): p. 228-235.

[17] Swami, A., *Impact of Automobile Induced Air Pollution on road side vegetation: A Review.* ESSENCE Int. J. Env. Rehab. Conserv. IX (1), 2018: p. 101-116.

[18] Gómez-Moreno, F.J., et al., *Urban vegetation and particle air pollution:*

Experimental campaigns in a traffic hotspot. Environmental Pollution, 2019. **247**: p. 195-205.

[19] Li, N., et al., *Pollution characteristics and human health risks of elements in road dust in Changchun, China.* International journal of environmental research and public health, 2018. **15**(9): p. 1843.

[20] Elad, D., et al., *In vitro exposure of nasal epithelial cells to atmospheric dust.* Biomechanics and Modeling in Mechanobiology, 2018. **17**(3): p. 891-901.

[21] Gasparotto, J., et al., *Obese rats are more vulnerable to inflammation, genotoxicity and oxidative stress induced by coal dust inhalation than non-obese rats.* Ecotoxicology and Environmental Safety, 2018. **165**: p. 44-51.

[22] Dong, L., et al., *Concentration and size distribution of total airborne microbes in hazy and foggy weather.* Science of The Total Environment, 2016. **541**: p. 1011-1018.

[23] Reich, P.B., et al., *Unexpected reversal of C3 versus C4 grass response to elevated CO2 during a 20-year field experiment.* Science, 2018. **360**(6386): p. 317-320.

[24] Vehviläinen, T., et al., *High indoor CO2 concentrations in an office environment increases the transcutaneous CO2 level and sleepiness during cognitive work.* Journal of Occupational and Environmental Hygiene, 2016. **13**(1): p. 19-29.

[25] Cetin, M., H. Sevik, and K. Isinkaralar, *Changes in the particulate matter and CO2 concentrations based on the time and weather conditions: the case of Kastamonu.* Oxidation Communications, 2017. **40**(1-II): p. 477-485.

[26] Yin, Y., et al., *Elevated CO2 levels increase the toxicity of ZnO nanoparticles to goldfish (Carassius auratus) in a water-sediment ecosystem.* Journal of Hazardous Materials, 2017. **327**: p. 64-70.

[27] Yu, T. and Y. Chen, *Effects of elevated carbon dioxide on environmental microbes and its mechanisms: A review.* Science of The Total Environment, 2019. **655**: p. 865-879.

[28] Swami, A., *Impact of Automobile Induced Air Pollution on roadside vegetation: A Review.* ESSENCE International Journal for Environmental Rehabilitation and Conservation, 2018. **9**(1): p. 101-116.

[29] Greenberg, N., et al., *Different effects of long-term exposures to SO2 and NO2 air pollutants on asthma severity in young adults.* Journal of Toxicology and Environmental Health, Part A, 2016. **79**(8): p. 342-351.

[30] Singh, B.K., A.K. Singh, and V.K. Singh, *EXPOSURE ASSESSMENT OF TRAFFIC-RELATED AIR POLLUTION ON HUMAN HEALTH-A CASE STUDY OF A METROPOLITAN CITY.* Environmental Engineering & Management Journal (EEMJ), 2018. **17**(2).

[31] Zhang, J., et al., *Sulfur dioxide inhalation lowers sperm quality and alters testicular histology via increasing expression of CREM and ACT proteins in rat testes.* Environmental Toxicology and Pharmacology, 2016. **47**: p. 47-52.

[32] Zhang, Q., et al., *The molecular mechanism of the effect of sulfur dioxide inhalation on the potassium and calcium ion channels in rat aortas.* Human & Experimental Toxicology, 2015. **35**(4): p. 418-427.

[33] Wookey, P.A., P. Ineson, and T.A. Mansfield, *Effects of atmospheric sulphur dioxide on microbial activity in decomposing forest litter.* Agriculture, Ecosystems & Environment, 1991. **33**(3): p. 263-280.

[34] Cassia, R., et al., *Climate Change and the Impact of Greenhouse Gasses: CO2 and NO, Friends and Foes of Plant Oxidative Stress.* Frontiers in Plant Science, 2018. **9**(273).

[35] Almaraz, M., et al., *Agriculture is a major source of NO<em>x</em> pollution in California.* Science Advances, 2018. **4**(1): p. eaao3477.

[36] Lu, X., et al., *Source apportionment and health effect of NOx over the Pearl River Delta region in southern China.* Environmental Pollution, 2016. **212**: p. 135-146.

[37] Salmón, P., et al., *Oxidative stress in birds along a NOx and urbanisation gradient: An interspecific approach.* Science of The Total Environment, 2018. **622-623**: p. 635-643.

[38] Feng, Z., et al., *Current ambient and elevated ozone effects on poplar: A global meta-analysis and response relationships.* Science of The Total Environment, 2019. **654**: p. 832-840.

[39] Nuvolone, D., D. Petri, and F. Voller, *The effects of ozone on human health.* Environmental Science and Pollution Research, 2018. **25**(9): p. 8074-8088.

[40] Kim, B.-G., et al., *Impact of ozone on claudins and tight junctions in the lungs.* Environmental Toxicology, 2018. **33**(7): p. 798-806.

[41] Alexopoulos, A., et al., *Experimental effect of ozone upon the microbial flora of commercially produced dairy fermented products.* International Journal of Food Microbiology, 2017. **246**: p. 5-11.

[42] Naidoo, G. and K. Naidoo, *Uptake of polycyclic aromatic hydrocarbons and their cellular effects in the mangrove Bruguiera gymnorrhiza.* Marine Pollution Bulletin, 2016. **113**(1): p. 193-199.

[43] Peterson, B.S., et al., *Effects of Prenatal Exposure to Air Pollutants (Polycyclic Aromatic Hydrocarbons) on the Development of Brain White Matter, Cognition, and Behavior in Later ChildhoodEffects of Prenatal Exposure to Air Pollutants on Children's BrainsEffects of Prenatal Exposure to Air Pollutants on Children's Brains.* JAMA Psychiatry, 2015. **72**(6): p. 531-540.

[44] Oyibo, J.N., et al., *Analysis of total petroleum hydrocarbons, polycyclic aromatic hydrocarbons and risk assessment of heavy metals in some selected finfishes at Forcados Terminal, Delta State, Nigeria.* Environmental Nanotechnology, Monitoring & Management, 2018. **9**: p. 128-135.

[45] Tiele, A., S. Esfahani, and J. Covington, *Design and Development of a Low-Cost, Portable Monitoring Device for Indoor Environment Quality.* Journal of Sensors, 2018. **2018**.

[46] Gugliermetti, L. and D. Astiaso Garcia, *A cheap and third-age-friendly home device for monitoring indoor air quality.* International Journal of Environmental Science and Technology, 2018. **15**(1): p. 185-198.

[47] Kumar, H., et al., *Solar Power Environmental Air Pollution & Water Quality Monitoring System Based on IoT.* Perspectives in Communication, Embedded-systems and Signal-processing-PiCES, 2018. **2**(8): p. 197-199.

[48] Saha, A.K., et al. *A raspberry Pi controlled cloud based air and sound pollution monitoring system with temperature and humidity sensing*. in *2018 IEEE 8th Annual Computing and Communication Workshop and Conference (CCWC)*. 2018.

[49] Yi, W.Y., et al., *A Survey of Wireless Sensor Network Based Air Pollution Monitoring Systems.* Sensors, 2015. **15**(12): p. 31392-31427.

[50] Levelt, P.F., et al., *The ozone monitoring instrument.* IEEE

Transactions on Geoscience and Remote Sensing, 2006. **44**(5): p. 1093-1101.

[51] Moltchanov, S., et al., *On the feasibility of measuring urban air pollution by wireless distributed sensor networks.* Science of The Total Environment, 2015. **502**: p. 537-547.

[52] Zoogman, P., et al., *Tropospheric emissions: Monitoring of pollution (TEMPO).* Journal of Quantitative Spectroscopy and Radiative Transfer, 2017. **186**: p. 17-39.

[53] Baron, R. and J. Saffell, *Amperometric Gas Sensors as a Low Cost Emerging Technology Platform for Air Quality Monitoring Applications: A Review.* ACS Sensors, 2017. **2**(11): p. 1553-1566.

[54] Kurien, C. and A.K. Srivastava. *Active Regeneration of Diesel Particulate Filter Using Microwave Energy for Exhaust Emission Control*. 2018. Singapore: Springer Singapore.

[55] Gysel, N., et al., *Particulate matter emissions and gaseous air toxic pollutants from commercial meat cooking operations.* Journal of Environmental Sciences, 2018. **65**: p. 162-170.

[56] Wang, G., et al., *Characteristics of particulate matter from four coal–fired power plants with low–low temperature electrostatic precipitator in China.* Science of The Total Environment, 2019. **662**: p. 455-461.

[57] Zhang, J. and Q. Mu, *Air pollution and defensive expenditures: Evidence from particulate-filtering facemasks.* Journal of Environmental Economics and Management, 2018. **92**: p. 517-536.

[58] Yu, S., et al., *Mitigation of severe urban haze pollution by a precision air pollution control approach.* Scientific Reports, 2018. **8**(1): p. 8151.

[59] Dey, S., et al., *Synthesis of silver promoted CuMnOx catalyst for ambient temperature oxidation of carbon monoxide.* Journal of Science: Advanced Materials and Devices, 2019.

[60] Ghosh, A., et al., *Catalytic oxidation and selective sensing of carbon monoxide for sense and shoot device using ZnO–CuO hybrids.* Materialia, 2019. **5**: p. 100177.

[61] Zinola, C.F., *Carbon monoxide oxidation assisted by interfacial oxygen-water layers.* Journal of Solid State Electrochemistry, 2019. **23**(3): p. 883-901.

[62] Osaka, Y., et al., *Basic study on exhaust gas purification by utilizing plasma assisted MnO2 filter for zero-emission diesel.* Separation and Purification Technology, 2019. **215**: p. 108-114.

[63] Lehtoranta, K., et al., *Particulate Mass and Nonvolatile Particle Number Emissions from Marine Engines Using Low-Sulfur Fuels, Natural Gas, or Scrubbers.* Environmental Science & Technology, 2019.

[64] Jiang, J. and D. Li, *Theoretical analysis and experimental confirmation of exhaust temperature control for diesel vehicle NOx emissions reduction.* Applied energy, 2016. **174**: p. 232-244.

[65] Roy, P. and A. Sardar, *SO^ sub 2^ Emission Control and Finding a Way Out to Produce Sulphuric Acid from Industrial SO^ sub 2^ Emission.* Journal of Chemical Engineering & Process Technology, 2015. **6**(2): p. 1.

[66] Zhong, Y., et al., *A model for performance optimization of wet flue gas desulfurization systems of power plants.* Fuel Processing Technology, 2008. **89**(11): p. 1025-1032.

[67] Evans, A.P., C. Miller, and S. Pouliot, *Operational experience of commercial, full scale ammonia-based wet FGD for over a decade.* Environmental Technologies, 2009: p. 1-19.

[68] Yang, D., et al. *Study on SO3 Cooperative Removal Effect of Ultra-low Emission Technology in Coal-fired Power Plants*. in *E3S Web of Conferences*. 2018. EDP Sciences.

[69] Cui, J., et al., *Effects of Air Pollution Control Devices on the Chlorine Emission from 410 t/h Circulating Fluidized Bed Boilers Co-firing Petroleum Coke and Coal.* Energy & Fuels, 2018. **32**(4): p. 4410-4416.

[70] Swallow, D., *EXHAUST SYSTEM WITH A MODIFIED LEAN NOx TRAP*. 2019, Google Patents.

[71] Wen, L., et al., *Summertime fine particulate nitrate pollution in the North China Plain: increasing trends, formation mechanisms and implications for control policy.* Atmos. Chem. Phys., 2018. **18**(15): p. 11261-11275.

[72] Shim, S.H., S.H. Jeong, and S.-S. Lee, *Low-nitrogen oxides combustion of dried sludge using a pilot-scale cyclone combustor with recirculation.* Journal of the Air & Waste Management Association, 2015. **65**(4): p. 413-422.

[73] Liu, H., et al., *Control of NOx emissions of a domestic/small-scale biomass pellet boiler by air staging.* Fuel, 2013. **103**: p. 792-798.

[74] Lin, F., et al., Flue gas treatment with ozone oxidation: an overview on NOx, organic pollutants, and mercury. Chemical Engineering Journal, 2020. p. 382.

[75] Rosenberg, H., et al., *Post combustion methods for control of NOx emissions.* Progress in Energy and Combustion Science, 1980. **6**(3): p. 287-302.

[76] Ahmad, T., et al., *Behavior of heavy metals in air pollution control devices of 2,400 kg/h municipal solid waste incinerator.* Korean Journal of Chemical Engineering, 2018. **35**(9): p. 1823-1828.

[77] Pei, S.-L., et al., *Performance evaluation of integrated air pollution control with alkaline waste valorization via high-gravity technology.* Journal of the Taiwan Institute of Chemical Engineers, 2018. **87**: p. 165-173.

[78] Revah, S. and J.M. Morgan-Sagastume, *Methods of odor and VOC control*, in *Biotechnology for odor and air pollution control*. 2005, Springer. p. 29-63.

[79] Moore, P.A., et al., *Development and Testing of the ARS Air Scrubber: A Device for Reducing Ammonia Emissions from Animal Rearing Facilities.* Frontiers in Sustainable Food Systems, 2018. **2**(23).

[80] Singh, S., Biosorption of Heavy Metals by Cyanobacteria: Potential of Live and Dead Cells in Bioremediation. In Microbial Bioremediation & Biodegradation 2020. p. 409-423. Springer, Singapore.

Chapter 6

Methane Cycling in Paddy Field: A Global Warming Issue

Mohammed Mahabubur Rahman and Akinori Yamamoto

Abstract

Paddy fields are major sources of CH_4 emission and a vital source of global warming. Thus, it is important to understand the CH_4 cycling in paddy field. The CH_4 chemistry, mechanisms of production and emission from paddy fields are also significantly important to understand. This paper discusses about the CH_4 cycling, how CH_4 emission effect on the global warming, and the mechanisms of CH_4 exchange between rice paddy field and atmosphere, factors effecting the CH_4 production, oxidation, transportation and calculation. Also try to suggest the CH_4 mitigation options of paddy fields. The mitigation of CH_4 emission can be achieved by water management, selection of rice cultivar and fertilization. Controlled irrigation can also reduce CH_4 production compared to flood irrigation. Cultivation of high-yielding and more heat-tolerant rice cultivars will be promising approach to reduce CH_4 emissions and slow down the global warming.

Keywords: paddy soil, methane cycling, atmosphere, mitigation, climate

1. Introduction

Methane (CH_4) is a key greenhouse gas (GHG), which has global warming potential 25-times higher than that of carbon dioxide (CO_2) over a 100-year period [1, 2]. Atmospheric CH_4 mixing ratio has increased approximately threefold since the pre-industrial era [3, 4]. The latest analysis of the World Data Centre for Greenhouse Gases (WDCGG) observations, the year-averaged CH_4 mixing ratio is 1853 ± 2 ppb in 2016. Atmospheric CH_4 mixing ratio has increased by 6.8 ppb y^{-1} over the last decade [5]. Paddy field is the dominant anthropogenic source of CH_4 [6–10], accounting for approximately 10% of global CH_4 emissions [11]. Emissions of CH_4 from paddy soils are concentrated in irrigated areas; irrigated paddy soils account for 60% of the total rice harvesting area worldwide, but produce 78% of CH_4 emissions in rice-producing areas [12]. Globally, around the tropics, sub-tropics and parts of the temperate boreal regions contribute most of the CH_4 emissions from paddy rice fields because these regions have a large area of paddy field compared with other regions. This includes areas like Central and Latin America, Africa and Southeast Asia [13]. Southeast Asia emits approximately 10 Mg CH_4 km^{-2} and it contributes 90% to the global rice CH_4 emissions chart [14]. Africa and South America contribute 3.5% and 4.7% to the global Paddy rice CH_4 budget respectively.

Rice is a vital crop in the world and it is grown on more than 167.25 million hectares of land globally [15]. In Asia, China, India, Indonesia, Bangladesh, and Vietnam are the major dominant rice-producing countries. So, the areas of high rice production are equally the area of high CH_4 emissions due to rice cultivation [14, 16].

Considering the large CH_4 emission from paddy fields, it is important to understand the CH_4 chemistry, mechanisms of CH_4 production and emission. Moreover, the reduction of CH_4 emission from rice paddy fields has become increasingly important. Therefore, the main objectives of this research are to discuss CH_4 cycling in the paddy soil and global warming, the basic understandings of methane chemistry, production, oxidation, transportation, calculation, the mechanisms of CH_4 exchange between rice paddy field and atmosphere, final emission and try to give some mitigation options of CH_4 emissions from paddy soils to slow down the global warming.

2. Rice cultivation, methane cycling and global warming

Global warming is a serious problem nowadays. CH_4 is one of the vital greenhouse gases that contribute to global warming. About 60% of global CH_4 emissions are occurred because of anthropogenic activities [17]. The main sources of anthropogenic CH_4 emissions are the oil and gas industries, agriculture, landfills, wastewater treatment, and emissions from coal mines. Globally, about 530 million tons of CH_4 (converted in terms of carbon) are emitted annually [17]. Rice fields are contributing to global warming, but it is a far bigger problem than previously thought. The conventional paddy field with continuous flooding irrigation is known as a major source of CH_4 emission [18]. The world's largest rice producers are China and India. After China and India, the next largest rice producers are Indonesia, Bangladesh, Vietnam, Myanmar, and Thailand. It is expected that economic activities will become more active in the next future mainly in Asia. Reportedly, of the Asian population of about 4200 million people, about 2700 million (60%) live on rice. Since Asia's population is expected to continue increasing, the total area of paddy fields in Asia will also increase. The paddy area in 2009 was 10,940,000 ha in Thailand, 4,410,000 ha in the Philippines, 7,420,000 ha in Vietnam and 12,100,000 ha in Indonesia [19]. So a hues amount of population needs more rice, more rice growing mean more CH_4 adding to the atmosphere. During 1 kg of rice grain production, paddy field contributes 100 g of CH_4 to the atmosphere. The default methane baseline emission factor is 1.3 kg CH4 $ha^{-1\text{-}day^{-1}}$, in continuous flooding rice cultivation [20, 21]. Part of the CH_4 produced in the rice soil is consumed in the oxidized rhizosphere of rice roots or in the oxidized soil-floodwater interface. Soil bacteria also can consume CH_4 [22]. CH_4 is also leached to ground water, as a small part dissolves in water and most of it escapes from the soil into the atmosphere (**see Figure 1**). The important thing is most of the CH_4 is staying in the atmosphere for 10 years as it is or the increasing CH_4 abundance leads to a longer lifetime for CH_4 [23]. After a certain time later, the CH_4 is broken down into CO_2. Approximately 95% of the CH_4 produced in flooded soils is oxidized to CO_2 before it release to the environment [24]. It is still not clear exactly how much of the CH_4 is finally converted to CO_2 and how much might remain as other intermediate carbon-containing compounds without a significant direct effect on the climate [25]. However, CH_4 initially reacts with ozone in a 'chain' reaction that ultimately produces CO_2 and water vapor. CH_4 also creates ground-level ozone in the atmosphere. And ozone is not only harmful to human health but also contributes to climate change. CH_4 is much more effective than CO_2 at trapping heat (more than 60%) in the atmosphere [26]. Global temperatures in 2014 and 2015 were warmer than at any other time in the modern temperature record after 1880. And carbon emissions are the central cause of that rise. Rising temperatures and changes in rainfall have a significant effect on enhancing microbial activity and create ideal conditions for microbial CH_4

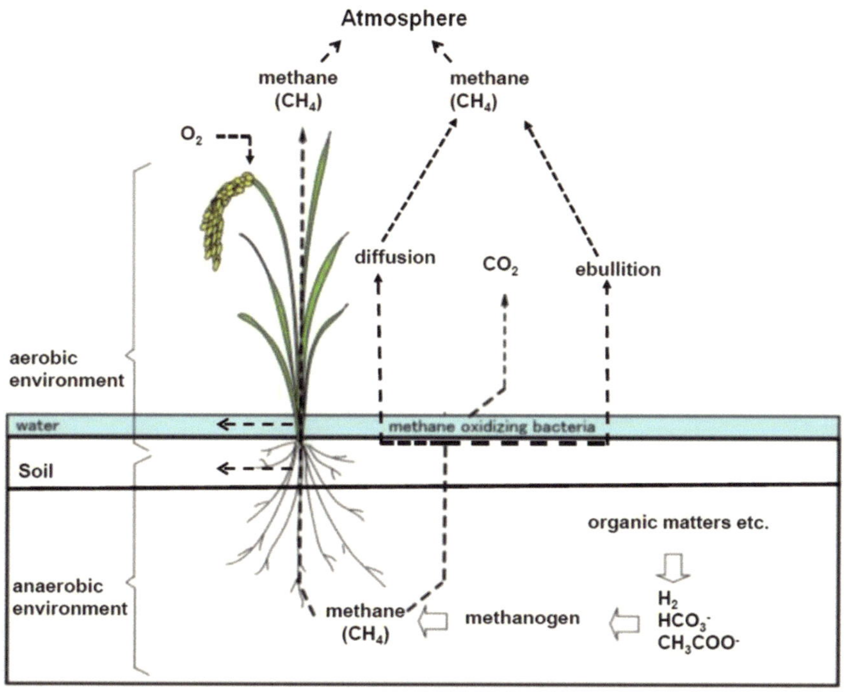

Figure 1.
Biogeochemical mechanisms of CH_4 production of paddy field and it's cycling (modify from [15]).

production of flooded rice fields. It is predicted that atmospheric temperature will rise 2°C by 2060. If it is actually going to happen, most of the agricultural plants cannot grow; it will make severe starvation for the world population. On the other hand, sea level will rise. The consequences will be so devastating, most of the low lands will go underneath the water and it will increase homeless people all over the world.

3. Methane chemistry

CH_4 is a very special kind of molecule. CH_4 is an end product of the organic carbon decomposition under anoxic conditions and the simplest organic compound and member of the paraffin series of hydrocarbons [27]. It is colorless, odorless gas that occurs abundantly in nature and as a product of anthropogenic activities. Its chemical formula is CH_4 (**Figure 2**). CH_4 is lighter than air, having a specific gravity of 0.554. It is slightly soluble gas in water and burns readily in air, forming carbon dioxide and water vapor; the flame is pale, luminous and very hot. The boiling point of CH_4 is –162°C and the melting point is –182.5°C. Basically, CH_4 is very stable, but mixtures of CH_4 and air, with the content between 5 and 14% by volume, are explosive [28].

Reactions of Methane

1. Combustion (oxidation)

   a. Complete oxidation: $CH_4 + 2\ O_2$, flame or spark $\rightarrow CO_2 + H_2O +$ heat

   b. Partial oxidation: $6\ CH_4 + O_2$, 1500°C $\rightarrow CO + H_2 + H_2C_2$
   $CH_4 + H_2O$, 850°, Ni $\rightarrow CO + H_2$

2. Halogenation
   $CH_4 + X_2$, heat → $CH_3X + HX$
   requires heat or light.
   $Cl_2 > Br_2$ no reaction with I_2.

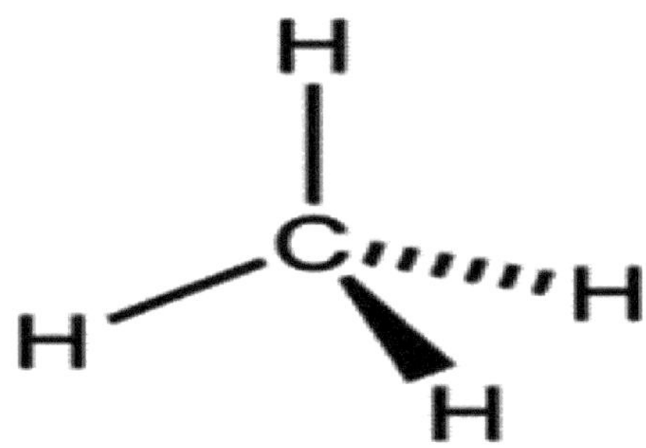

Figure 2.
Chemical formula of methane.

4. Biogeochemical mechanisms of methane production of paddy fields

The production of CH_4 is a microbiological process, which is predominantly controlled by the absence of oxygen and the amount of easily degradable actions [29]. Methanogens produce CH_4 under anaerobic conditions [30, 31]. Methanogens are prokaryotic microorganisms and belong to the domain of archaea. They are living in an anaerobic environment (e.g., soil or water) or in the intestines of animals. Methanogens mainly use acetate (contributes about 80% to CH_4 production) as a carbon substrate but another substrate like H_2/CO_2 and formats also contribute 10–30% to CH_4 production [32]. Acetate and hydrogen are formed by fermentation from hydrolyzed organic matter [29]. However, flooding of rice fields cuts off oxygen supply from the atmosphere to the soil, which leads to anaerobic fermentation of organic matter in the soil, resulting in the production of CH_4 [33]. And thereafter much of it escapes from the soil into the atmosphere via gas spaces in the rice roots and stems, and the remainder CH_4 bubbles up from the soil and/or diffuses slowly through the soil and overlying flood water (**Figure 1**). In flooded rice paddies, straw incorporation usually stimulates CH_4 production [34]. Root exudates and degrading roots are also important sources of CH_4 production, especially at the later growth stages of paddy. There are two major pathways of CH_4 production (e.g., acetoclastic and hydrogenotrophic). Acetoclastic methanogens use ATP to convert acetate to acetyl phosphate and then remove the phosphate ion via a reaction catalyzed by coenzyme A [12]. CH_4 is formed gradually by processes involving oxidized ferredoxin, tetrahydrosarcinapterin, coenzyme M, and coenzymes B. Taking account of all, CH_4 emissions from paddy soil are the net result of CH_4 production, oxidation and transportation. The complete CH_4 production process can be expressed as reduction and oxidation of two molecules of a simple hydrocarbon, one of which is reduced to CH_4 and the other of which is oxidized to CO_2: $2CH_2O \rightarrow CO_2 + CH_4$ [15].

5. Methane transportation from paddy soil to atmosphere

The total CH_4 emission process consists of three ways from soil to atmosphere e.g., transport via rice plants; bubble ebullition and molecular diffusion through the paddy water (see **Figure 1**). CH_4 can escape from the rice paddy soil via aerenchyma in the plant (90%), ebullition (10%), and diffusion through the soil and water layer (1%). CH_4 transports via the plant starts in the roots; CH_4 enters by diffusion through the epidermis and during the water uptake. It is likely that dissolved CH_4 is directly gasified in the root cortex and further diffuses upwards to the root-shoot transition zone traveling through intercellular spaces and aerenchyma. The aerenchyma system is developed by the plant to transport the oxygen necessary for respiration from leaves towards the roots. Just like CH_4 diffuses from the soil into the root system, oxygen diffuses from the root into the soil, creating a relative oxygen-rich zone in the rhizosphere. CH_4 is partly oxidized in the rhizosphere to CO_2 by methanotrophic bacteria. Methanogenesis in the rhizosphere itself is suppressed by oxygen. The transport of CH_4 to the atmosphere depends on the properties of the rice plant. The flux of gases in the aerenchyma depends on permeability coefficients, concentration gradients of roots and the internal structure of the aerenchyma. The number of tillers m^{-2}, root mass, rooting pattern, total biomass, and metabolic activity also influence gas fluxes.

6. Factors effecting the methane production in paddy fields

6.1 Effects of temperature on methane production

Temperature is one of the major determining factors on the biological process (e.g., within the soil), which controls the CH_4 production. Previous studies showed that increased soil temperature leads to an increase in CH_4 production [35]. There is a lot of qualitative evidence showing that CH_4 production from rice field increase with the increasing temperature [35]. A laboratory experiment regarding CH_4 production from two rice cultures incubated at temperatures between 20 and 38°C showed E_α values of 41 and 53 kJ mol^{-1}. CH_4 production in anoxic paddy soil suspensions incubated between 7 and 43°C showed E_α values between 53 and 132 kJ mol^{-1} with an average value of 85 kJ mol^{-1} [36]. The paddy soil temperature could control the amount of CH_4 production and there is a positive and strong correlation in both soil temperature and CH_4 production pattern [37]. The effect of temperature on CH_4 production in paddy soil was investigated by [38], they found that in continuously flooded soil, the temperature optimum for CH_4 production was 40°C, however, this shifted to 45°C during a period of intermittent irrigation accompanied by a marked decrease in activity. The optimum temperature during the non-cropping season was also 45°C (**Figure 3**).

6.2 Effects of soil pH and Eh on methane production in paddy field

Soil pH is another influential factor in CH_4 production. The pH effects on CH_4 production in a flooded rice soil. Methanogenic bacteria are acid sensitive. Generally, the optimum pH for methanogenesis is 7.0. Introducing the acidic materials frequently results in a decrease in CH_4 production [39]. A slight decrease in soil pH can cause decreases in CH_4 production. A slight increase in soil pH (about 0.2 unit higher than the natural soil suspension pH) resulted in an enhancement of CH_4 production by 11 to 20% and 24 to 25% at controlled Eh of −250 and − 200 mV, respectively [40]. These results suggested that a small reduction of soil pH could be obtained a decrease in CH_4 production in paddy soil.

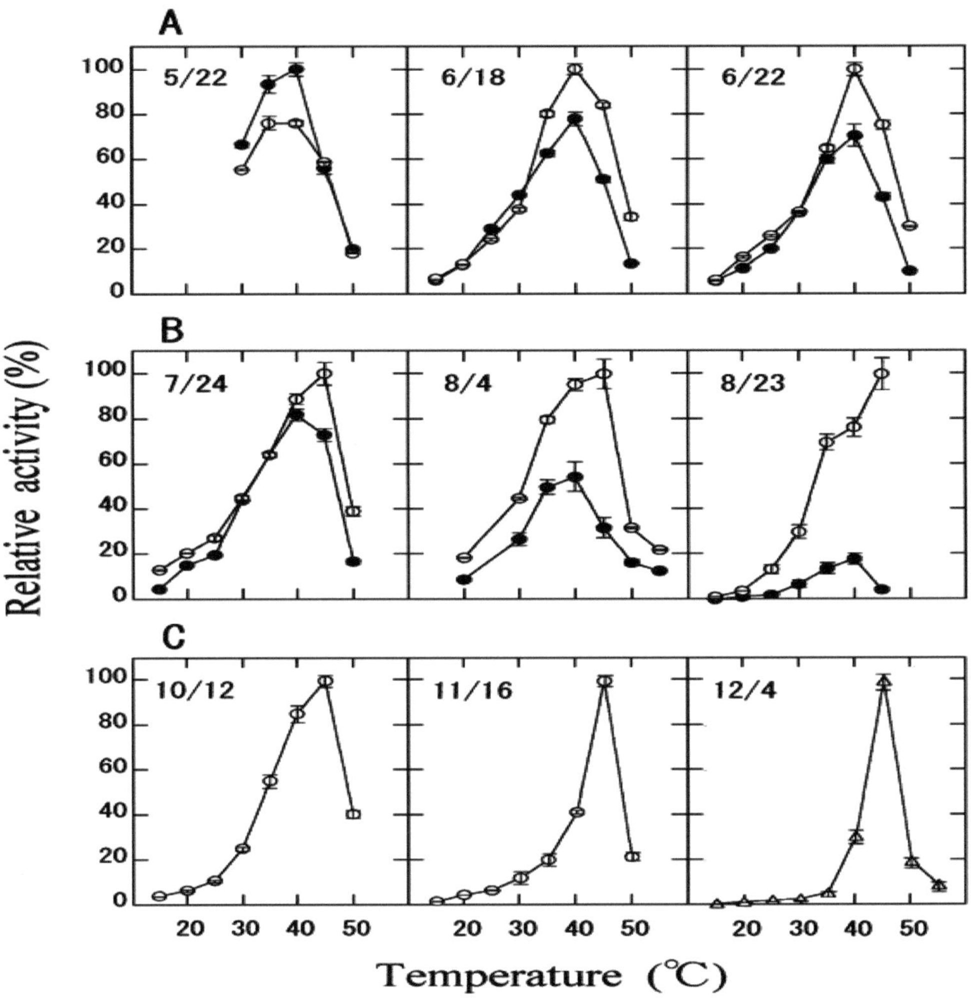

Figure 3.
Effects of temperature on CH_4 production in paddy soil shown in each figure. Periods of soil sampling: (A) continuous flooding (B) intermittent irrigation (C) from harvest to winter [38].

6.3 Effects of paddy cultivar, root and rhizosphere on methane production

The decomposition of organic matter by methanogens under anaerobic condition leads to the production of CH_4. Rice cultivars have significant effects on CH_4 production of planted soil due to large variation in composition and content of root exudates [41]. The roots of rice plants are colonized by methanogenic Archaea, which performs the last step in the production and emission of CH_4 [42–45]. It has previously been shown using pulse labeling of rice plants with $^{13}CO_2$ that plant photosynthesis accounts for more than 50% of total CH_4 emission [45, 46] and that a particular group of methanogens, the Rice Cluster I is responsible for the production of CH_4 from rice photosynthates [47]. It is well established that the type of rice cultivar affects the CH_4 production from rice fields [48]. In the randomly determined CH_4 production experiment found the initial production of CH_4 was much larger on roots grown in the rice field (RF) than in low-level river bank soil (LL), as shown by factorial ANOVA ($P < 0.0001$), while the effect of the different rice cultivars was not significant (P = 5 0.7401). The randomly determined CH_4 production rates during the initial 3–8 days activity of the methanogenic community are shown in **Figure 4a** [49]. The rates of CH_4 production also increased

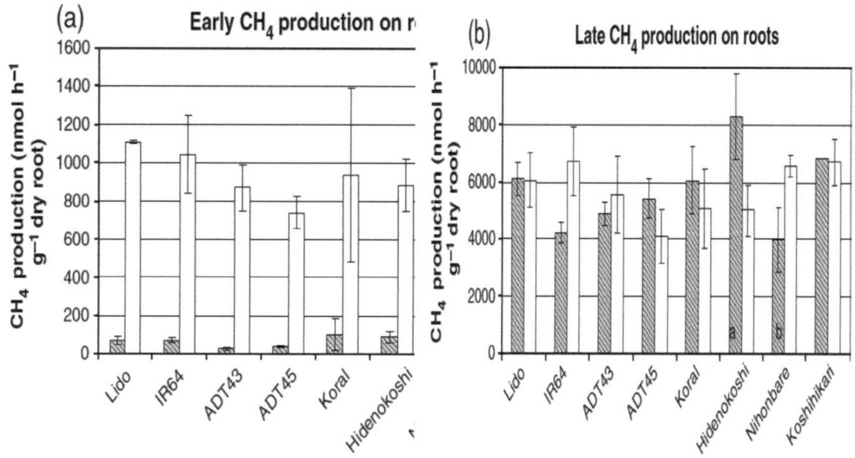

Figure 4.
Rates of CH_4 production on excised roots from eight different rice cultivars retrieved from microcosms consisting of RF soil (open bars) or LL soil (hatched bars). The rates of CH_4 production were determined (a) after 3–8 days of incubation and (b) after 17–28 days of incubation [49].

with incubation time, those of roots grown on LL soil after a prolonged lag phase, but in a late phase, arbitrarily chosen after 17–28 days, reached a similar value in all different root samples assayed (**Figure 4b**) [49]. The root morphophysiological traits were positively and significantly correlated with grain yield, whereas root length, specific root length, root oxidation activity, root total and active absorbing surface area were negatively and significantly correlated with total CH_4 emission [10]. However, rice plants can enhance CH_4 production by providing substrates for methanogenesis through the production of root litter and root exudates that contain carbohydrates and amino acids [50]. These nutrients stimulate microbial activities and lead to an increase in CH_4 production [51, 52]. In terms of the substrates, root exudates, especially in the form of acetate, represented a considerable source of CH_4 production at the rice ripening stage [53]. The CH_4 emission peak occurring at the rice ripening stage because this stage attributed to the easily decomposable organic matter exuded from roots [54]. The rhizospheric CH_4 oxidation at the different rice growth stages in detail. Under laboratory conditions, 29% of CH_4 oxidation in the rhizosphere was found one week before panicle initiation and no CH_4 oxidation occurred at the rice ripening stage. The CH_4 oxidation in the rhizosphere at the harvest stage may be negligible in the field [55]. The rhizospheric CH_4 oxidation rate varied with rice growing stages, being lower at the tillering stage (36.5% of CH_4 produced in rice rhizosphere) than at the panicle initiation stage (54.7%). This may be related to the oxidizing activity of rice roots varying with the rice growth stage [50]. At the late tillering stage, root exudates dominate CH_4 production of planted soil [41, 56]. This would probably be due to the stimulating effect of plant roots on O_2 released during the decomposition of soil organic carbon exceeding CH_4 oxidation in the rice rhizosphere [41, 57]. The relative contribution of root-associated CH_4 production to CH_4 emissions could be important in the rice paddies, as it varied between 4 and 52% [45].

6.4 Effects of paddy growing stages on methane emission

More than 90% of the total CH_4 emitted during the cropping season of the rice plants. Emission through rice plants may be expected to show great seasonal variations as a function of changes in soil conditions and variations in plant growth. The CH_4 emissions varied during the growth period of the rice plant. The CH_4 emission low during the early growth stages of the rice plant [58]. This may be due to low levels of methanogenesis in this stage. And the high amounts of CH_4 emissions were measured during the reproductive and ripening stages. This was happened probably due to the higher availability of fatty acids and sugars in this stage [58]. A significantly high amount of CH_4 emission occurred at the reproductive stage of the rice plants in the paddy field (**Figure 5**) [59].

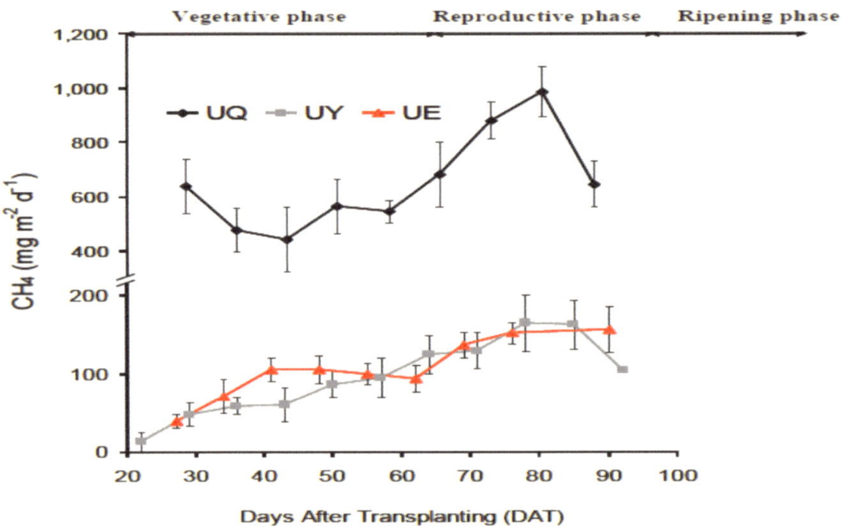

Figure 5.
CH_4 emission pattern of different stages of rice growth [59].

6.5 Effects of field management (fertilization, water management, farming operations) on methane production

In the rice paddy filed, organic fertilizer application (such as animal manure, sewage sludge, crop residues) significantly enhances soil nutrient availability, microbial biodiversity and their activity [60]. Consequently, the increased availability of carbon after the application of organic fertilizer increases CH_4 emissions and leads to a clear shift in the dominant methanogens in paddy soil [2, 61]. The CH_4 emission and soil fertility both were significantly increased by multiyear organic fertilization in paddy soil [62]. The **Figure 6** shows the effects of organic fertilizer on the pathway of CH_4 production from rice soil. In the rice production, water management and nitrogen (N) fertilizers are the two main driving factors of greenhouse gas emissions [63]. N application can increase the production of CH_4 in paddy fields. The level of the microbial community and NH^{+4} stimulates the growth and activity of CH_4 oxidation bacteria, reducing CH_4 efflux [64]. N fertilizers might also affect CH_4 production at the level of the microbial community [65]. In terms of water management, controlled irrigation can reduce CH_4 production compared to flood irrigation. Flooding conditions in the fields cause limited oxygen and other gasses such as sulfates in that soil environment. This condition promotes methanogenesis activities that release more CH_4 emission to the atmosphere [66]. Field burning of agricultural residues also results in release of CH_4, nitrous oxide and other minor GHGs. Rice straw is a perspicuous type of organic matter to apply in terms of the carbon cycle in paddy field. The rice straw increased CH_4 emission significantly [67].

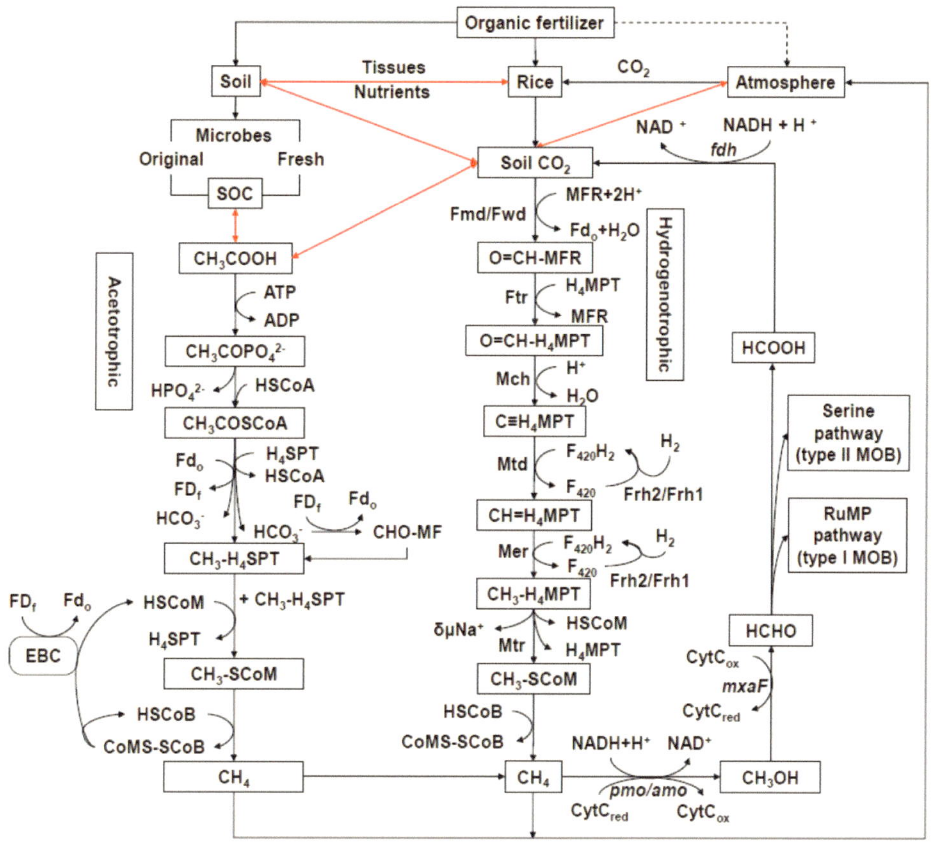

Figure 6.
Effects of organic fertilizer on the pathway of CH_4 production from rice soil [12].

7. Oxidation process of methane in the paddy field soil

In paddy fields, CH_4 production and oxidation happen simultaneously, so it is difficult to directly determine CH_4 production and oxidation separately. CH_4 oxidation may occur in aerobic and anaerobic ways. Previously CH_4 oxidation was usually determined by comparing CH_4 fluxes from flooded soils under aerobic and anaerobic incubation conditions [68]. By using this approach, CH_4 oxidation accounted for up to >90% of CH_4 production [52]. With the development of CH_4 oxidation inhibitors and isotopic methods, CH_4 oxidation is now widely measured using these approaches and shows relatively low ratio (less than 70%) to CH_4 production [69]. Generally, CH_4 emission is extremely influenced by the balance between CH_4 production and CH_4 oxidation in paddy fields [70]. Oxidation of CH_4 reduces the emission of CH_4 in the soil of rice field to the atmosphere. CH_4 is relatively inert in anoxic environments, but is oxidized by methanotrophic bacteria as soon as oxygen becomes available [71]. Aerobic methanotrophic bacteria are present in the oxic surface layer of the submerged paddy soil and in the rhizosphere where oxygen is available in a shallow layer around the rice roots. The most distinctly possible sites for CH_4 oxidation in rice fields are the water–soil interface and the rhizosphere. It has been shown that CH_4 oxidation is taking place within these zones of the paddy soil so that part of the produced CH_4, does not reach the atmosphere [72]. The increase of CH_4 production may, in turn, stimulate methanotrophic growth and CH_4 oxidation [73], but methanotrophic growth can be limited by low oxygen concentrations [74, 75]. Because larger rice plants stimulate oxygen transport into the rhizosphere, water management practices can also affect CH_4 oxidation rates through their effect on plant growth. The anaerobic CH_4 oxidation in the subsoil of a rice paddy was below 5% of the CH_4 emission during 38 whole growth periods [76]. The microbial aerobic oxidation activity of CH_4, the population of aerobic CH_4 oxidizers and the factors influencing the activity of CH_4 oxidation were investigated in three types of paddy rice soil in Zhejiang Province, China. The experiment results concluded that the population of CH_4 oxidizing bacteria was at maximum within the peak-tillering, heading and flowering stages, during which the largest population of methanogenic bacteria also appeared. Temperatures from 25 to 35°C and pH from 6 to 8 were the optimum conditions for aerobic oxidation of CH_4 in paddy rice soil [77]. Soil particle sizes also significantly affect the activity of CH_4 oxidation. Approximately 95% of the CH_4 produced in flooded soils is oxidized to carbon dioxide before it is released to the environment. Therefore, the oxidation is important for the biogeochemical cycling of CH_4 [24].

8. Measurements techniques of methane emission from paddy field

Recently, scientists are applying several techniques for measuring the CH_4 emission from the paddy field.

8.1 Closed chamber method

The most common technique for measuring the CH_4 emission in the rice paddy field is the closed chamber method (**Figure 7**) [24]. Normally, the chamber made of plexiglass (size: 50 cm length × 50 cm width × 100 cm height). The chamber equipped with a fan to make sure the inside gas mixing during chamber deployment. The chamber basement equipped with a water seal. Gas sampling needs to do simultaneously at three points per plot once a week. Normally, gas samples collect at 0, 5, 10, 20, and 30 min after the chamber deployment between 7:00–10:00 am. The samples taken by a syringe and store in evacuated glass vials and then subjected to the laboratory analysis using gas chromatography. The best hour for representing average daily flux is when temperatures are close to the daily mean, i.e., at 10 a.m. This is the best way to estimate the daily cumulative value from a unique measurement of the day [67]. The main advantages of this method are detecting low fluxes, of being easy to manipulate, low costs and flexibility [78].

8.2 Eddy covariance method using an open-path gas analyzer

The eddy covariance method is a complex, expensive and advanced method for measuring the CH_4 emission from the real-life rice paddies (**Figure 8**). The eddy covariance method calculates fluxes of a scalar of interest (i.e. CH_4, CO_2, LE, and H) at the same time, measuring turbulent fluctuations in vertical wind speed and the scalar and then computing the covariance between the two [79]. A sonic anemometer-thermometer (CSAT3) measures three-dimensional wind speed and sonic temperature. An open-path CO_2/H_2O gas analyzer (LI-7500A) measures fluctuations in CO_2 and water vapor densities and an open-path CH_4 analyzer (LI-7700) measures CH_4 concentrations [79]. CSAT3, the point of reference, has to fix at certain meters above the ground. The data from CSAT3, LI-7500A, and LI-7700 sampled at 10 Hz using a data logger (CR3000). The eddy covariance raw data need to process and quality control use EddyPro software to compute the fluxes of CH_4, CO_2, LE, and H over a 30-minute interval.

Eddy covariance can give a better picture of how much CH_4 real-life rice paddies are emitting, for example, detect and quantify CH_4 from rice paddies [80]. Compare

Figure 7.
The closed chamber technique of CH_4 sampling in rice planted paddy field.

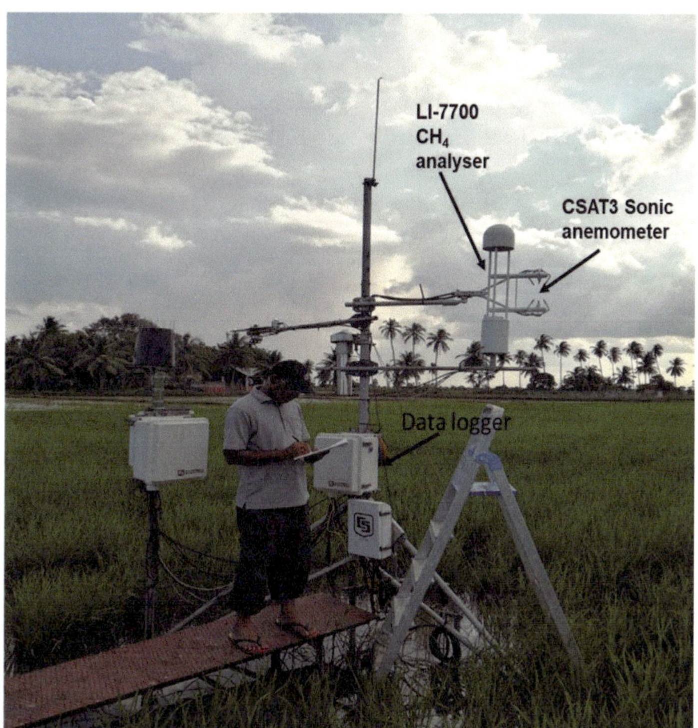

Figure 8.
Set-up of eddy covariance system with CSAT3 sonic anemometer, LI-7500A open-path CO_2/H_2O analyzer, and LI-7700 open-path CH_4 analyzer.

with an open-path and a closed path CH_4 analyzer to provide advances in understanding the performance and limitations of the eddy covariance method applied to CH_4 measurements, from an instrumental and flux processing point of view. Closed-path CH_4 analyzers require high flow rates for significantly reduced optical cell pressures to provide adequate response time and sharpen absorption features [81]. Closed path CH_4 analyzer can detect low fluxes and it is less expensive and flexible, on the other hand, the eddy covariance method using an open-path gas analyzer is highly sensitive, expensive and complex but gets a better estimation of CH_4 flax from paddies [82].

8.3 The calculation of the total methane emission (E_{CH4})

The CH_4 can be calculated by following Equation (1).

$$^{E}CH_4\left(=E_{plant}+E_{bubble}+E_{diffusion}\right) \quad (1)$$

At first, E_{plant} is given by.

$$E_{plant}=k_p\times k_{tp}\times\int_{root}\times LAI\times C_{CH4soil}\cdot \quad (2)$$

Where k_p is the turnover rate of the methane emission via rice plant (=0.03 day^{-1}), $\int_{tp}$ is a factor of CH_4 emission defined for each paddy type (=15.0), $\int_{root}$ is a distribution factor of rice root in the soil, and LAI is the leaf area index ($m^2\ m^{-2)}$.

E_{bubble} is given by.

$$E_{bubble}=k_b\times(C_{CH4soil}-C_{thresh}), \quad (3)$$

Where k_b is a turnover rate of the CH_4 emission as bubble (=1.0 day^{-1}), and C_{thresh} is the dissolved CH_4 threshold at which CH_4 bubble formation occurs (=6.0 gC m^{-3}).

$E_{diffusion}$ is given by,

$$E_{diffusion}=(C_{CH4soil}-C_{CH4atm})\times\int_{dif}\times\int_{tor}\times p_{soil}, \quad (4)$$

Where C_{CH4} atm is the atmospheric CH_4 concentration (=1.0 × 10–3 gC m-3), $\int_{dif}$ is a diffusion coefficient of CH_4 (=1.73 × 10^{-4} m^2 day^{-1}), $\int_{tor}$ is a tortuosity coefficient of the soil (=0.66), and p_{soil} is the soil porosity defied for each soil type (mm mm^{-1}).

9. Mitigation of methane emission from paddy field

CH_4 emission increases during flooding condition and decrease when water drain from the field [83]. Therefore, the irrigation system is one of the most vital tools of rice farming and is the most important effort for CH_4 mitigation. Thus, the water level is one of the main factors affect the production and emission of CH_4 in rice paddy fields. In irrigated rice, a short aeration period at tillering has been shown to increase yield. Single mid-season drainage may reduce seasonal emission rates by about 50% [52]. In a controlled experiment, CH_4 emissions can be significantly reduced when the field is drained and dried at mid tillering and before harvest [53]. Higher yielding rice genotypes could be viable options for reducing the release of CH_4 from paddy fields to the atmosphere [13]. System of Rice Intensification (SRI) is the alternative rice farming for climate change adaptation and mitigates greenhouse gas emission from paddy fields. The study showed that the SRI paddy field with intermittent wetting-drying irrigation reduced CH_4 emission by up to 32% [84]. SRI application can save water irrigation up to 28%, 38.5% and 40% in Japan, Iraq and Indonesia, respectively [85]. SRI also reduced greenhouse gas emission that is represented by reducing global warming potential up to 37.5% in Indonesia [86] and 40% in India [87].

The activity of CH_4 producing bacteria is inhibited from the oxidizing condition of paddy soil by water management. CH_4 emission factor for intermittently flooded paddy fields can decrease by approximately 20% [88]. Water management and organic material management are significant for reducing CH_4 emissions from rice paddy fields. Mid-season drainage and intermittent flooding are effective for increasing the productivity and quality of rice as well as reducing CH_4 emissions. Mitigation of CH_4 emissions from rice paddy fields by water management has huge potential to be marketed. A field experiment in the Philippines showed that direct seeding techniques reduced CH_4 emissions by 18% as compared with transplanted rice. Another study, in Japan, showed that global warming potential declined by 42% just by changing the puddle of rice seedlings to zero tillage. Moreover, CH_4 emission can be reduced by shifting to more heat-tolerant rice cultivars and by adjusting sowing dates. It will also prevent yield decline due to temperature increases [89]. A multi-criteria evaluation ranking score for CH_4 mitigation strategy is been done in Egypt. They found that short duration rice varieties got the highest score. This strategy could be reduced 25% of CH_4 emission, reduce 20% of water consumption without any reduction in rice productivity. On the other hand, fertilizer management strategy was in the second-ranking and followed by the midseason drainage [90]. The heat-tolerant improved rice variety with low CH_4 emission is a potential mitigation option [91].

10. Conclusions

The most important aim of studies on CH_4 emission from paddy fields is the mitigation of global warming and adaptation with climate change. CH_4 emission is controlled by several factors such as temperature, soil pH, Eh, rice cultivar, root, rhizosphere, group of methanogens, paddy growing stages and field management.

From the above-mentioned discussion, it is clear how CH_4 cycling in paddy fields, also how much CH_4 release to atmosphere and leached to ground water and soil.

CH_4 can be reduced from paddy field through management practices like, the mid-season drainage, alternate wetting, drying irrigation and using alternative fertilizers have been shown to reduce CH_4 emissions from rice paddies and achieve sustain rice production.

By using a combination of feasible mitigation technologies there is great potential to reduce CH_4 emission from rice fields and increase rice production. Crop improvement strategies such as breed high yielding and high stress tolerant rice varieties with reduced CH_4 emissions will help the CH_4 mitigation. These rice varieties should be adaptable to changing climate e.g., the stress and water management conditions.

Cultivation of high-yielding and more heat-tolerant rice cultivars will be a promising approach to reduce CH_4 emissions from paddy fields and slowing down global warming.

Future research needs to focus on global paddy field CH_4 budgets, climate change adaptation policy and sustainable agriculture technology, creating more disease and heat resistance rice variety.

Acknowledgements

We are thankful to journal editors and authors for giving permission to reuse the figures in this manuscript. This research was supported by the Environment Research and Technology Development Fund (2-1502 and 2-1802) of the Environmental Restoration and Conservation Agency of Japan. We are also grateful to Mr. Michel Rahman, the University of Toronto for his painstaking proofreading.

Conflict of interest

The authors declare no conflict of interest.

Author details

Mohammed Mahabubur Rahman* and Akinori Yamamoto
Natural Science Research Unit, Tokyo Gakugei University, 4-1-1 Nukuikitamachi, Koganei, Tokyo, Japan

*Address all correspondence to: mmrahman@u-gakugei.ac.jp

References

[1] Yvon-Durocher G, Allen AP, Bastviken D, Conrad R, Gudasz C, St-Pierre A, Thanh-Duc N, del Giorgio PA. Methane fluxes show consistent temperature dependence across microbial to ecosystem scales. Nature. 2014; 507:488-491. DOI: 10.1038/nature13164

[2] Yuan J, Yuan YK, Zhu YH, Cao LK. Effects of different fertilizers on methane emissions and methanogenic community structures in paddy rhizosphere soil. Sci Total Environ. 2018; 627:770-781. DOI: 10.1016/j.scitotenv.2018.01.233

[3] Zou J, Huang Y, Jiang J. A 3-year field measurement of methane and nitrous oxide emissions from rice paddies in China: effects of water regime, crop residue, and fertilizer application. Global Biogeochemical Cycles. 2005; 19. DOI:10.1029/ 2004GB002401. GB2021.

[4] Keppler F, Hamilton JTG, Braß M, Rockmann T. Methane emissions from terrestrial plants under aerobic conditions. Nature. 2006; 439:187-191, DOI:10.1038/nature04420.

[5] WMO [Internet]. 2018. WMO WDCGG data summery, https://gaw.kishou.go.jp/static/publications/summary/sum42/sum42.pd [Accessed: 2020-01-07]

[6] Ehhalt DH, Schmidt U. Sources and Sinks of atmospheric methane. Pure and Applied Geophysics. 1978; 116:452-464

[7] Khalil MAK, Rasmussen RA. Sources, sinks, and seasonal cycles of atmospheric methane. J Geophys Res. 1983; 88:5131-5144. DOI:10.1029/JC088iC09p05131

[8] Tian, H. Q., et al. The terrestrial biosphere as a net source of greenhouse gases to the atmosphere. Nature. 2016; 531(7593):225-228. DOI: 10.1038/nature16946

[9] Fumoto, T. [Internet]. 2016. Process-Based Modeling of Methane Emissions from Rice Fields.https://www.naro.affrc.go.jp/publicity_report/pub2016_or_later/files/niaes_report38-2.pdf [Accessed: 2020-02-22]

[10] Wang J, Akiyama H, Yagi K, Yan X. Controlling variables and emission factors of methane from global rice fields. Atmospheric Chemistry and Physics. 2018; 18: 10419-10431. DOI:10.5194/acp-18- 10419-2018

[11] Ma K, Conrad R, Lu Y. Responses of methanogen mcrA genes and their transcripts to an alternate dry/wet cycle of paddy field soil. Applied and Environmental Microbiology. 2012;78:445-454. doi: 10.1128/AEM.06934-11

[12] Yuan J, Yi X and Cao L, 2019. Three-Source Partitioning of Methane Emissions from Paddy Soil: Linkage to Methanogenic Community Structure. International Journal of Molecular Sciences. 2019; 20:1586. DOI:10.3390/ijms20071586

[13] Wang C Lai DYF, Sardans J, Wang WQ, Zeng CS, Penuelas J. Factors ~ related with CH4 and N2O emissions from a paddy field: clues for management implications. Plos One. 2017;12:e0169254

[14] FAO [Internet]. 2020. FAOSTAT Emissions Database, Agriculture, Rice Cultivation. Available from: http://www.fao.org/faostat/en/#data/GR [Accessed 2020/09/11]

[15] Tokida T, Fumoto T, Cheng W, Matsunami T, Adachi M, Katayanagi N, Matsushima M, Okawara Y, Nakamura H, Okada M, Sameshima R, Hasegawa T. Effects of

free-air CO2 enrichment (FACE) and soil warming on CH4 emission from a rice paddy field: impact assessment and stoichiometric evaluation. Biogeosciences. 2010; 7:2639-2653. Doi:10.5194/ bg-7-2639-2010

[16] Yan X Y, Akiyama H, Yagi K, Akimoto H. Global estimations of the inventory and mitigation potential of methane emissions from rice cultivation conducted using the 2006 Intergovernmental Panel on Climate Change Guidelines. Global Biogeochemcal Cycle. 2009; 23:GB2002. DOI:2010.1029/2008GB003299

[17] Ito K. 2015. Suppression of Methane Gas Emission from Paddy Fields, Nippon Steel and Sumitomo Metal Technical Report, No. 19, pp. 145-148.

[18] Cicerone RJ, Delwiche CC, Tyler SC, Zimmerman PR. Methane emissions from California rice paddies with varied treatments. Global Biogeochemical Cycles. 1992; 6:233-248. DOI:10.1029/92GB01412

[19] Statista [Internet]. 2019. https://www.statista.com/statistics/271969/world-rice-acreage-since-2008/ [Accessed: 2019-12-12]

[20] IPCC [Internet]. 2006. IPCC Guidelines for National Greenhouse Gas Inventories, prepared by: National Greenhouse Gas Inventories Programme. https://www.ipcc-nggip.iges.or.jp/public/2006gl/pdf/0_Overview/V0_0_Cover.pdf [Accessed 2020/09/23]

[21] Balakrishnan D, Kulkarni K, Latha PC, Subrahmanyam D. 2018. Crop improvement strategies for mitigation of methane emissions from rice, Emirates Journal of Food and Agriculture. 30(6): 451-462

[22] Conrad R. Mechanisms controlling methane emission from wetland rice fields. In: Oremland RS, editor. The Biogeochemistry of Global Change: Radiative Trace Gases. Chapman & Hall; 1993. p. 317-335.

[23] Prather MJ. Time-scales in atmospheric chemistry: theory for GWPs for CH_4 and CO, and runaway growth. Geophysical Research Letters. 1996; 23:2597-2600. DOI:10.1029/96GL02371

[24] Wassmann R, Buendia LV, Lantin RS, Bueno CS, Lubigan LA, Umali A, Nocon N, Javellana AM, Neue HU. Mechanisms of crop management impact on methane emissions from rice fields in Los Baños, Philippines. Nutrient Cycling in Agroecosystems. 2000. 58: 107-119. DOI: 10.1023/A:1009838401699

[25] Boucher OP, Friedlingstein B. Collins and K. Shine. The indirect global warming potential and global temperature change potential due to methane oxidation. Environmental Research Letters. 2009. 4:044007. doi:10.1088/1748-9326/4/4/044007.

[26] Shindell D, Faluvegi G. Climate response to regional radiative forcing during the twentieth century. Nature Geoscience. 2009; 2:294-300.

[27] Conrad R. Control of microbial methane production in wetland rice fields. Nutrient Cycling in Agroecosystems. 2002; 64:59-69. DOI: 10.1023/A:1021178713988

[28] Britannica [Internet] 2019. Metnahe, https://www.britannica.com/science/methane. [Accessed 2019/12/13]

[29] Segers R. Methane production and methane consumption: A review of processes underlying wetland methane fluxes. Biogeochemistry. 1998; 41, 23-51. DOI: 10.1023/A:1005929032764

[30] Svensson BH, Sundh I. Factors affecting methane producton in peat soils. Suo. 1992; 43:183-190.

[31] McAllister SA, Kral TA. Methane production by methanogens following an aerobic washing procedure: simplifying methods for manipulation. Astrobiology. 2006; 6: 819-823. DOI: 10.1089/ast.2006.6.819

[32] Chin KJ, Conrad R. Intermediary metabolism in methanogenic paddy soil and the influence of temperature. FEMS Microbiology Ecology.1995;18: 85-102. DOI: 10.1111/j.1574-6941.1995.tb00166.x

[33] Ferry JG. Methane from acetate. Journal of Bacteriology. 1992;174:5489-5495. DOI: 10.1128/jb.174.17.5489-5495.1992

[34] Lu Y, Wassmann R, Neue HU, Huang C. Dynamics of dissolved organic carbon and methane emissions in a flooded rice soil. Soil Science Society of America Journal. 2000. 64: 2011-2017. DOI: 10.2136/sssaj2000.6462011x

[35] Khalil MAK, Rasmussen RA, Shearer MJ, Chen ZL, Yao H, Yang J. Emissions of methane, nitrous oxide, and other trace gases from rice fields in China. Journal of Geophysical Research Atmospheres.1998; 103:25241-25250. DOI: 10.1029/98JD01114

[36] Schütz H, Seiler W, Conrad R. Influence of soil temperature on methane emission from rice paddy fields. Biogeochemistry.1990;11:77-95. DOI: 10.1007/BF00002060

[37] Fazli P, Hasfalina CM, Mohamed Azwan M Z, Umi Kalsom MS, Nor Aini AR, Azni I. Methane emission from paddy soil in relation to soil temperature in tropical region. Jurnal Teknologi. 2016. 78: 87-91. OI: 10.11113/jt.v78.7273

[38] Hattori C, Ueki A, Seto T, Ueki K. Seasonal Variations in Temperature Dependence of Methane Production in Paddy Soil. Microbes and Environments. 2001; 16.:227-233. DOI:10.1264/jsme2.2001.227.

[39] Alexander M. Introduction to soil microbiology. 2nd ed. John Wiley & Sons, New York. 1977. 467p.DOI: 10.1177/003072707901000208

[40] Wang ZP, DeLaune RD, Masscheleyn PH, Patrick WH Jr. Soil Redox and pH Effects on Methane Production in a Flooded Rice Soil. Soil Science Society of America Journal. 2009; 57:382-385. DOI: 10.2136/sssaj1993.03615995005700020016x

[41] Setyanto P, Rosenani AB, Boer R, Fauziah CI, Khanif MJ. The effect of rice cultivars on methane emission from irrigated rice field. Indonesian Journal of Agricultural Science. 2004; 5, 20-31. DOI: 10.21082/ijas.v5n1.2004.20-31

[42] Grosskopf R, Stubner S, Liesack W. Novel euryarchaeotal lineages detected on rice roots and in the anoxic bulk soil of flooded rice microcosms. Applied and Environmental Microbiology. 1998; 64: 4983-4989

[43] Chin K J, Lueders T, Friedrich M W, Klose M, Conrad R. Archaeal community structure and pathway of methane formation on rice roots. Microbial Ecology. 2004;47, 59-67. doi: 10.1007/s00248-003-2014-7

[44] Lu YH, Conrad R. In situ stable isotope probing of methanogenic archaea in the rice rhizosphere. Science. 2005; 309: 1088-1090. DOI: 10.1126/science.1113435

[45] Minoda T, Kimura M, Wada E. Photosynthates as dominant source of CH_4 and CO_2 in soil water and CH_4 emitted to the atmosphere from paddy fields. Journal of Geophysical Research. 1996;101:21091-21097. DOI: 10.1029/96JD01710

[46] Watanabe A, Takeda T, Kimura M. Evaluation of origins of CH_4 carbon emitted from rice paddies. Journal of Geophysical Research. 1999; 1042(D19):23623-23630. DOI: 10.1029/1999JD900467

[47] Conrad R, Klose M. Dynamics of the methanogenic archaeal community in anoxic rice soil upon addition of straw. European Journal of Soil Science. 2006; 57:476-484. doi: 10.1111/j.1365-2389.2006.00791.x

[48] Aulakh MS, Bodenbender J, Wassmann R, Rennenberg H. Methane transport capacity of rice plants. I. Influence of methane concentration and growth stage analyzed with an automated measuring system. Nutrient Cycling in Agroecosystems. 2000;58: 357-366. DOI:10.1023/A:1009831712602

[49] Conrad R, Klose M, Noll M, Kemnitz D, Bodelier PLE. Soil type links microbial colonization of rice rootsto methane emission. Global Change Biology. 2008; 14: 657-669. DOI: 10.1111/j.1365-2486.2007.01516.x

[50] Jia Z, Cai Z, Xu H, Li X. Effect of rice plants on CH_4 production, transport, oxidation and emission in rice paddy soil. Plant and Soil. 2001; 230:211-221. DOI: 10.1023/A:1010366631538

[51] Holzapfel-Pschorn A, Conrad R, Seiler W. Effects of vegetation on the emission of methane from submerged paddy soil. Plant and Soil. 1986;92: 223-233. DOI: 10.1007/BF02372636

[52] Sass RL, Fisher FM, Wang YB, Turner FT, Jund, MF. Methane emission from rice fields: The effect of flood Water management. Global Biogeochemical Cycles, 1992;6:249-262. DOI: 10.1029/92GB01674

[53] Neue HU, Wassmann R, Lantin RS, Alberto MCR, Aduna JB, Javellana AM. Factors affecting methane emission from rice fields. Atmospheric Environment. 1996; 30: 1751-1754. DOI: 10.1016/1352-2310(95)00375-4

[54] Wang B, Neue HU, Samonte HP. Effect of cultivar differences (IR 72, IR 65598, and Dular) on methane emission. Agriculture, Ecosystems and Environment. 1997;62: 31-40.

[55] Van der Gon HACD, Neue HU. Oxidation of methane in the rhizosphere of rice plants. Biology and Fertility of Soils. 1996; 22: 359-366. DOI: 10.1007/BF00334584

[56] Watanabe I, Hashimoto T, Shimoyama. A. Methane oxidizing activities and methanothropic populations associated with wetland rice plants. Biology and Fertility of Soils.1997; 24: 261-265. DOI: 10.1007/s003740050241

[57] Jia ZJ, Cai ZC, Xu H, Tsurata H. Effects of rice cultivars on methane flux in a paddy soils. Nutrient Cycling in Agroecosystems. 2002; 64: 87-94. DOI: 10.1023/A:1021102915805

[58] Bharati K, Mohanty SR, Rao VR, Adhya TK. Influence of flooded and non-flooded conditions on methane efflux from two soils planted to rice. Chemosphere. 2001; 3:25-32. DOI:10.1016/S1465-9972(00)00034-9

[59] Gaihre YK, Tirol-Padre A, Wassmann R, Aquino E, Pangga VG, Sta Cruz PC. . Spatial and temporal variations in methane fluxes from irrigated lowland rice fields. Phllipine Agricultural Science. 2011;94: 335-342.

[60] Jannoura R, Joergensena RG, Bruns C. Organic fertilizer effects on growth, crop yield, and soil microbial biomass indices in sole and intercropped peas and oats under organic farming conditions. European Journal of Agronomy. 2014;52, 259-270. DOI:10.1016/j.eja.2013.09.001

[61] Zhou B, Wang Y, Feng Y, Lin X. The application of rapidly composted manure decreases paddy CH_4 emission by adversely influencing methanogenic archaeal community:a greenhouse study. Journal of Soils and Sediments. 2016; 16: 1889-1900.

[62] Yuan J, Sha ZM, Hassani D, Zhao Z, Cao LK. Assessing environmental impacts of organic and inorganic fertilizer on daily and seasonal greenhouse gases effluxes in rice field. Atmos. Environ. 2017; 155:119-128. DOI: 10.1016/j.atmosenv.2017.02.007.

[63] Nie TZ, Chen P, Zhang ZX, Qi, ZJ, Lin YY, Xu D. Effects of different types of water and nitrogen fertilizer management on greenhouse gas emissions, yield, and water consumption of paddy fields in cold region of China. International Journal of Environmental Research and Public Health. 2019;16: 1639. DOI: 10.3390/ijerph16091639

[64] Schimel J. Global change: Rice, microbes and methane. Nature. 2000; 403, 375-377.

[65] Cai Z, Shan Y, Xu H. Effects of nitrogen fertilization on CH4 emissions from rice fields. Soil Science and Plant Nutrition. 2007; 53:353-361.

[66] Bouwman AF. Exchange of greenhouse gases between terrestrial ecosystems and the atmosphere. In: Bouwman AF, editor. Soils and the Greenhouse Effect, Chichester:Willy; 1990; p 61-127.

[67] Minamikawa K, Sakai N, Yagi K. Methane emission from paddy fields and its mitigation options on a field scale. Microbes and Environments. 2006;21:135-147. DOI: 10.1264/jsme2.21.135

[68] Conrad R, Rothfuss F. Methane oxidation in the soil surface layer of a flooded rice field and the effect of ammonium. Biology and Fertility of Soils. 1991;12, 28-32. DOI: 10.1007/BF00369384

[69] Kruger M, Eller G, Conrad R, Frenzel P. Seasonal variation in pathways of CH_4 production and in CH_4 oxidation in rice fields determined by stable carbon isotopes and specific inhibitors. Global Change Biology. 2002; 8: 265-280. DOI: 10.1046/j.1365-2486.2002.00476.x

[70] Zhao X, He J, Cao J. Study on mitigation strategies of methane emission from rice paddies in the implementation of ecological agriculture. Energy Procedia. 2011; 5: 2474-248. DOI: 10.1016/j.egypro.2011.03.425

[71] Rudd JWM, Taylor CD. Methane cycling in aquatic environments, In: Droop MR, Jannasch MR, editors. Advances in Aquatic Microbiology. London. 1980. p.77-150.DOI: 10.1002/jobm.3630210813

[72] Holzapfel-Pschorn A, Conrad R, Seiler W. Production, oxidation and emission of methane in rice paddies. FEMS Microbiology Letters. 1985;31:343-351. DOI: 10.1016/0378-1097(85)90030-8

[73] Cai Y, Zheng Y, Bodelier PLE, Conrad R, Jia Z. Conventional methanotrophs are responsible for atmospheric methane oxidation in paddy soils. Nature Communications. 2016; 7:11728. DOI: 10.1038/ncomms11728

[74] Conrad R. Microbial ecology of methanogens and methanotrophs. Advances in Agronomy. 2007; 96:1-63. DOI: 10.1016/S0065-2113(07)96005-8

[75] Jiang Y, Qian H, Huang S, Zhang X, Wang L, Zhang L, Shen M, Xiao X, Chen F, Zhang H, Lu C, Li C, Zhang J, Deng A, van Groenigen KJ, Zhang W. Acclimation of methane emissions from rice paddy fields to straw addition. Science Advances. 2019;5:eaau9038. DOI: 10.1126/sciadv.aau9038

[76] Murase J, Kimura M. Methane production and its fate in paddy fields. 9. Methane flux distribution and decomposition of methane in

the subsoil during the growth period of rice plants. Soil Science and Plant Nutrition.1996;42: 187-190. DOI: 10.1080/00380768.1996.10414702

[77] Min H, Chen ZY, Wu WX, Chen MC. Microbial aerobic oxidation of methane in pady soil. Nutrient Cycling in Agroecosystems. 2002;64:79-85. DOI: 10.1023/A:1021127621257

[78] Wassmann R, Alberto MC, Tirol-Padre A, Hoang NT, Romasanta R, Centeno CA, Sander BO. Increasing sensitivity of methane emission measurements in rice through deployment of "closed chambers" at nighttime. PLoS ONE, 2018;13(2):e0191352. DOI: 10.1371/journal.pone.0191352

[79] Alberto MCR, Wassmann R, Buresh RJ, Quilty JR, Correa TQ Jr, Sandro JM, Centeno CAR. Measuring methane flux from irrigated rice fields by eddy covariance method using open-path gas analyzer. Field Crops Research. 2014; 160:12-21. DOI:10.1016/j.fcr.2014.02.008

[80] Meijide A, Manca G, Goded I, Magliulo V, di Tommasi P, Seufert G, Cescatti A. Seasonal trends and environmental controls of methane emissions in a rice paddy field in Northern Italy. Biogeosciences Discussions. 2011; 8:3809-3821. DOI: 10.5194/bgd-8-8999-2011

[81] McDermitt D, Burba G, Xu L, Anderson T, Komissarov A, Reinsche B, Schedlbauer, J, Starr G, Zona D, Oechel W, Oberbauer S, Hastings S. A new low-power, open-path instrument for measuring methane flux by eddy covariance. Applied Physics B. 102, 391-405. DOI: 10.1007/s00340-010-4307-0

[82] Walter BP, Heimann M. A process-based, climate-sensitive model to derive methane emissions from natural wetlands: Application to five wetland sites, sensitivity to model parameters, and climate. Glob. Biogeochem. Cycles, 2000; 14:745-765. doi:10.1029/1999GB001204

[83] Setyanto P, Makarim AK, Fagi AM, Wassmann R, Buendia LV. Crop management affecting methane emissions from irrigated and rainfed rice in Central Java (Indonesia). Nutrient Cycling in Agroecosystems. 2000; 8:85-93. DOI:10.1007/978-94-010-0898-3_8

[84] Rajkishore SK, Doraisamy P, Subramanian KS, Maheswari M. Methane Emission Patterns and Their Associated Soil Microflora with SRI and Conventional Systems of Rice Cultivation in Tamil Nadu, India. Taiwan Water Conservancy. 2013; 61: 126-134.

[85] Hameed KA, Mosa AKJ, Jaber FA. Irrigation water reduction using System of Rice Intensification compared with conventional cultivation methods in Iraq. Paddy and Water Environment. 2011;9 :121-127. DOI: 10.1007/s10333-010-0243-1

[86] Hidayah S, Agustina DA, Joubert MD, Soekrasno. [internet]. 2009. Intermittent irrigation in System of Rice intensification potential as an adaptation and mitigation option of negative impacts of rice cultivation in irrigated paddy field. http://www.rid.go.th/thaicid/_6_activity/Technical-Session/SubTheme2/2.10-Susi_H-Dewi_AA-Marasi_DJ-Soekrasno.pdf [Accessed: 2020-09-17]

[87] Gathorne-Hardy A, Reddy DN, Venkatanarayana M, Harriss-White B. A Life Cycle Assessment (LCA) of greenhouse gas emissions from SRI and flooded rice production in SE India. Taiwan Water Conservancy. 2013; 61: 110-125.

[88] Nagata A. [Internet]. 2010. Mitigation of Methane Emissions from Rice Paddy Fields in Japan, Ministry

of Agriculture, Forestry and Fisheries, Japan https://globalmethane.org/expo-docs/india10/postexpo/ag_nagata.pdf [Accessed: 2020-02-15]

[89] Van Groenigen KJ, Van Kessel C, Hungate BA. Increased greenhouse-gas intensity of rice production under future atmospheric conditions. Nature Climate Change. 2013; 3, 288-291. DOI: 10.1038/nclimate1712

[90] Hasan E. Proposing mitigation strategies for reducing the impact of rice cultivation on climate change in Egypt. Water Science. 2013; 27:69-77. DOI: 10.1016/j.wsj.2013.12.007

[91] Balakrishnan D, Kulkarni K, Latha PC, Subrahmanyam D. Crop improvement strategies for mitigation of methane emissions from rice. Emirates Journal of Food and Agriculture. 2018;30: 451-462. DOI: 10.9755/ejfa.2018.v30.i6.1707

Chapter 7

A Revisit of Rainfall Simulator as a Potential Tool for Hydrological Research

Felix Gemlack Ngasoh, Constantine Crown Mbajiorgu, Matthew Boniface Kamai and Gideon Onyekachi Okoro

Abstract

Different means of hydrological data collection have developed and used. However, they are constraint in one way or other. This paper therefore revisited the rainfall simulator as potential tool for hydrological research. The research disclosed that there are three different types of rainfall simulators; drop former simulator, pressure nozzle simulator and hybrid simulator. It can further be classified as indoor model and outdoor. The research also showed that precipitation is the driving force in hydrological studies. Consequently, in the design of rainfall simulator, the following should be taken into consideration: nozzle spacing, pump size, nozzle size, nozzle type, nozzle spacing, plot size and pressure. Meanwhile, intensity, distribution uniformity, kinetic energy, rainfall drop size and rainfall terminal velocity should be noted in its evaluation. Factoring-in the aforementioned design considerations, data collection is made easy without necessarily waiting for the natural rainfall. Since the rainfall can be controlled, the erratic and unpredictable changeability of natural rainfall is eliminated. Emanating from the findings, pressurized rainfall simulator produces rainfall characteristics similar to natural rainfall, which is therefore recommended for laboratory use if natural rainfall-like characteristics is the main target.

Keywords: rainfall-simulator, intensity, uniformity, kinetic-energy, drop-size, runoff, hydrology, research-tool

1. Introduction

Disintegration of the soil are impelled by the effect of rain drops on plain or almost plain soils, which detaches and splash soil particles and transports them downslope as a feature of surface flow. The net disintegration rate (sediments mass/unit zone) is an element of both rain sprinkle and surface flow. Runoff from earth surface conveys with it the most erodible sediment and fine sand particles from the dirt surface as the water streams downhill. At that point, rills are shaped; they start little channels, inevitably framing gaps, which can bring about enormous soil losses [1]. These processes are regularly studied in the field with normal precipitation which may be moderated by uncontrolled factors such as irregularity of the precipitation events. This paper therefore reviewed rainfall simulator as a potential tool for hydrological research.

2. Rainfall simulator

Rainfall simulators (RS) are device designed to model the characteristics of natural rainfall to the nearest possible. It can be used to determine inter-rill erosion rates and their dependent rainfall and soil parameters [1]. It has been a tool for agricultural research and has been used for different studies ranging from determination of soil characteristic, such as infiltration rate, surface runoff, storage or erosion process studies [2]. Yakubu and Yusop [3] pointed out two most important aspects to note while using rainfall simulator; the method used to simulating rainfall and runoff from plot. Consideration was not given to infiltration.

3. Rainfall simulator classifications

There are three classification of rainfall simulator: drip, pressurised nozzle (PN) [3–5] and hybrid [6] rainfall simulator.

Drip simulator: also known as drop former (DF) [4] uses hanging yarn or hypodermic needles to produce drops of necessary size at zero velocity. Its impact velocity is attained by free fall which made others defined it as non-pressurised simulator [7]. The drilled holes and drop height determines the diameter of the raindrop and kinetic energy respectively [3] (see **Figure 1**).

It is capable of producing drops which ranges from 3 to 6 mm depending on the diameter holes [8]. Main advantage of the drip simulator is that it has the ability to produce relatively large drops at low application rate [5, 9]. It has the following disadvantages [10]:

i. It is impractical for field since it requires huge distance of at least 10 m height to attain terminal velocity.

Figure 1.
Drop formers simulator [11].

ii. Another constraint of this simulator is that simulation is only carried out on a limited plot depending on the size of the hanging yarn.

iii. It does not produce distribution drops unless a variety of drop forming sized tubes are used.

Pressurised Simulator (PN): produces drop distribution that includes both small and large range of drop sizes with nonzero initial velocity and an impact velocity similar to terminal velocity of raindrops. In order to obtain drops of suitable sizes while upholding high velocity, high discharge nozzles are required [7]. The application rates are reduced by means of an intermittent moving object which intercepts the rainfall. An example of this type of simulator was developed by [12]. The authors found out that it utilises the best nozzle known as "yet-for-rain simulation" (spraying system 80,100-veejet nozzle). But problem with the 80,100-veejet nozzles was that it did not simulate rainfall energy characteristic and is still better than other nozzles. This type of rainfall simulator provides about 80% of the required kinetic energy per volume of natural rain [5, 12, 13]. This nonzero pressurised nozzle has the following advantages over the hanging yarn simulator [10] as presented by **Figure 2**:

i. They can be used in the field and their intensities can be varied more than the drop forming type of simulator.

ii. Since the nozzle simulator has an initial velocity greater zero, it requires shorter height to reproduce the terminal velocity obtained from natural rain.

iii. According to Home, (2017), this simulator is often portable compared to drop former.

Hybrid type simulator: uses the principles of pressurised and drip former techniques of simulation incorporated together. It was first developed by [15] to reduce the kinetic energy impact of the rainfall, but the research indicated that the technique reduced the kinetic energy at the detriment of the rain uniformity [3]. Wildhaber et al. used a similar method by placing mesh 0.5 m of aperture 2 mm × 1.7 mm under a spraying nozzle. The obtained result was not far from the non-pressurised simulator type. Carvalho et al. also designed a pressurised nozzle simulator with mesh placed 2.35 m below the nozzle to change rainfall characteristics (see **Figure 3**), and varying the nozzles and mesh types. The results varied based on the aperture of the meshes employed. Conclusively, the hybrid simulator was noted as suitable tool to assessing erodibility of different types of soil.

According to its transportability, rainfall simulators are classified as indoor and outdoor [7].

Indoor rainfall simulator: this rainfall simulator is used for modelling precipitation in a controlled environment. It is also known as Laboratory scale model. This simulator reduces lot of disadvantages incurred by the transportable type of rainfall simulator [7]. For example, Darboux et al. designed an indoor rainfall simulator system that simulated infiltration, run-off and erosion (see **Figure 4**), and the output was effective but the system was constrained with lack of non-recycle of the water system as well as not portable.

Outdoor rainfall simulator: could be portable or large depending on the projected purpose. Many of these types of simulator have been used to relate soil surface characteristics and controlling to runoff, infiltration and erosion as influenced by different parameters [18–22].

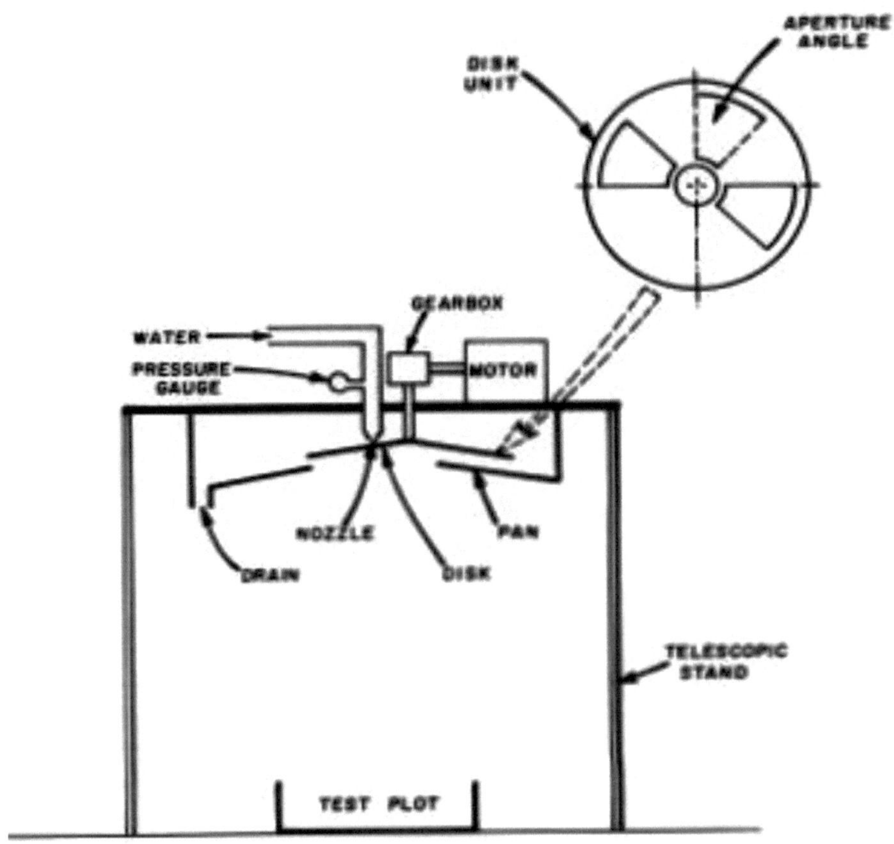

Figure 2.
Pressurised nozzle simulator [14].

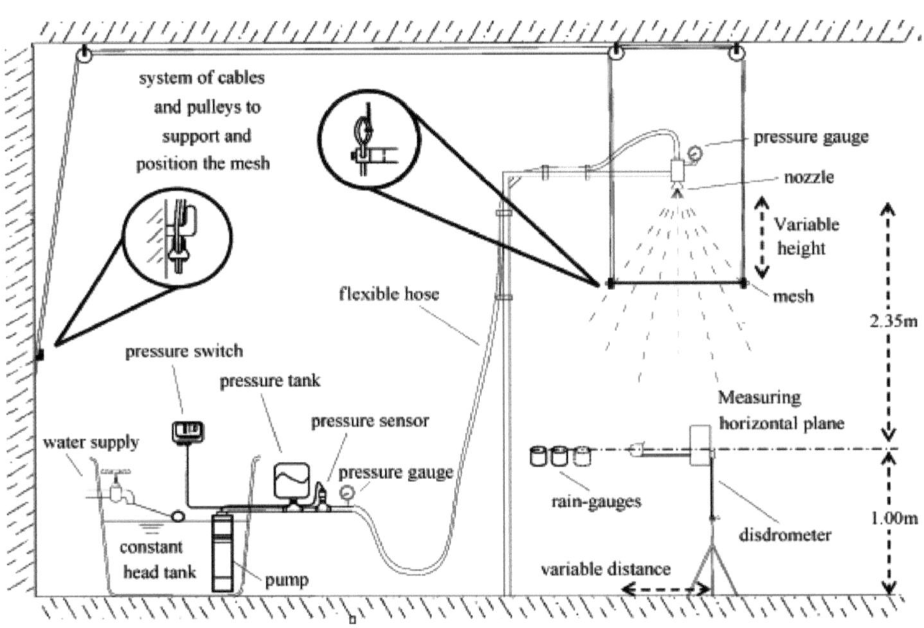

Figure 3.
Hybrid simulator [16].

A research carried out at Duke University Durham, using large transportable rainfall simulator of area 15.12 m^2 (**Figure 5**). The system was tested with common pressure washing nozzles which produced rainfall intensity of 62.43 mm/h and 32 mm h^{-1} with a corresponding uniformity coefficient (C.U) of 76.65 and 62% [13]. [23] developed a portable field simulator for use in hillside and obtained a consistent raindrop size of 2.58 mm with an intensities of 20 to 90 mm h^{-1} and C.U of 91.7% at an intensity of 60 mmh^{-1}. In a similar event Abudi et al. [24] also designed and constructed a portable rainfall simulator for field investigation of runoff, the drop size obtained was 1.5 mm with a ground hitting velocity near that of natural rainfall and energy flux 76% of the natural rainfall. All the simulators offered good

Figure 4.
Rainfall simulation building [17].

Figure 5.
Outdoor rainfall simulator [13].

performance. The merit of these simulators is that it can be used to study field parameters required for hydrologic modelling on any surface including the ones covered with vegetation. But they were limited by problem of natural rainfall which resulted to dismantling the setup when experiment schedule was not over and water was not recycled.

From the numerous studies carried out on the simulation of rainfall both in field and laboratory experiment, two merits of rainfall simulator in a research carried out in 2010 using laboratory simulator [25] were pointed out as:

i. It is faster in data collection without waiting for the natural rain.

ii. With rainfall simulator, you can work with controlled rain, thereby, eliminating the erratic and unpredictable changeability of natural rainfall.

4. Requirements and considerations for rainfall simulator

The characteristics of raindrop is also important for storm-water management purpose particularly in relation to understanding runoff process [26, 27]. Rainfall simulation should exemplify the following fundamental characteristic of the natural rain [7, 28].

i. Drop size distribution

ii. Terminal velocity

iii. Distribution uniformity

iv. The rainfall intensity and

v. Kinetic energy

4.1 Drop size distribution

One of the basic natural rainfall characteristics that are considered is it drop size which ranges between 0.5 and 5 mm [3]. The measurement of rain droplets sizes has been studied using various approaches [28], but there is no established standard for obtaining raindrop diameter size [3]. Basically, there are two methods used for determining drop size; manual and automatic raindrop measurement [26].

The manual measurement techniques of drop size distribution includes; stain, flour pellet, oil immersion and photographic methods while automated raindrop measurement techniques include; impact disdrometers (acoustic and displacement); optical disdrometer (optical image and optical scattering). **Figure 6** presents a tree of the drop size distribution classification. Drop size can be determined using Eq. (1):

$$D_r = \sqrt[3]{\frac{6}{\pi}W} \quad (1)$$

where W is the weight of the formed.

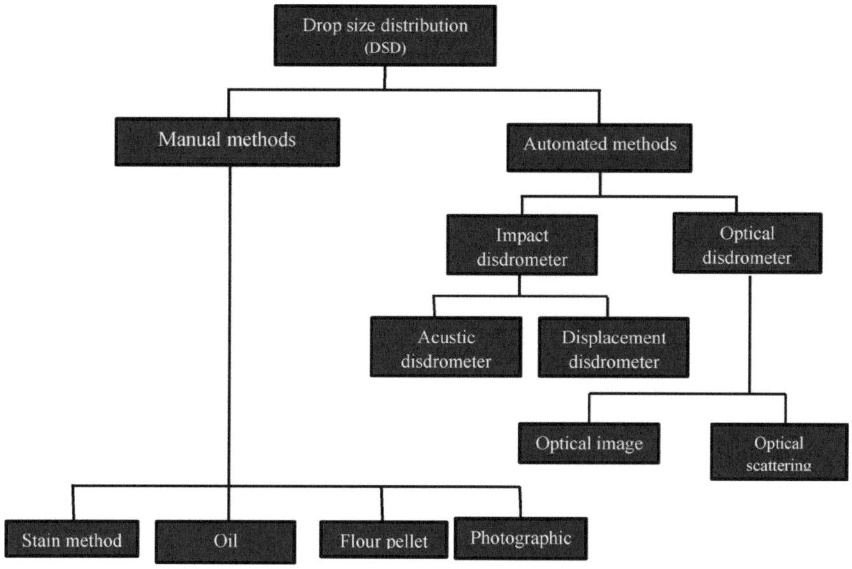

Figure 6.
Classification of drop size distribution methods.

4.2 Terminal velocity

A natural raindrop from greater height tends to reach terminal velocity before impact. This impact produces several effects on soil disintegration and infiltration. This is important particularly for studying soil erosion challenges where drops should reach their terminal velocity before impact [28]. This rainfall characteristic highly depends on the height of the simulator [3]. When the downward gravitational forces acting on the rainfall are cancelled out by the drag acting on the drop, the terminal velocity is achieved. Terminal velocity have been measured by many researchers using electronic devices to estimate the time for drops to pass consecutive point via photograph during fall [1, 7, 22, 28, 29] by stopwatch, timing the individual fall from a known height or simple computation [28, 30]. Computation of velocity of drop reaching the ground at an angle in natural precipitation (storm) with wind conditions 3. Simulated rainfall if well done can attain up to 94% of the terminal velocity of the natural rainfall [3, 31].

4.3 Rain intensity

One of the major ways to assess rainfall simulators is by the simulated rain intensity which the means by which other rainfall characteristics are defined, especially the rain impact kinetic energy [3]. Another characteristic that correlated with intensity is the drop size distribution [11]. The method to control rain intensity varies in rainfall simulator. But it is quite a difficult task most especially using drop forming simulator because it involved the manual movement of the frame [3, 11, 32]. In the case of pressurised nozzle simulators, intensity and drop diameter are control the varying the pressure [3] or by introducing a body in a swinging or rotating motion under the nozzle [6, 14].

4.4 Kinetic energy

Kinetic energy of rainfall is the degree by which the energy of the rain is measured. It is the major factor in soil detachment process. The energy of the rain is relational to its "erosivity" [1], and it is expressed in Jm^{-2} mm^{-1}. The technique of varying kinetic energy differs among rainfall simulators and the purpose for which a research is carried out [3]. Obtaining higher kinetic energy with drop forming simulator is an indication of the non-portability of the simulator because it requires higher height get such KE value. Aksoy et al. [33] in an investigation obtained kinetic energy of 21 Jm^{-2} mm^{-1}, using pressurised nozzle simulator at lower rainfall intensity of 45 mm/h and a height of 2.4 m. By varying the drop diameter from 2.7 to 5.1 mm and height of fall from 0.17 to 2.5 m, similar result was obtained [34].

The kinetic energy of rainfall is depending on two factors; terminal velocity at impact and the spraying nozzle which give intensity. Therefore when a simulator is designed for investigation of potential erosion by simulated rainfall, these aforementioned two factors should be taken note of [29]. This can however, be determined by using Eq. (2) [35].

$$\mathrm{KE} = 0.119 + 0.0873 \log \mathrm{I} \tag{2}$$

where KE is the kinetic energy of the rainfall in ($MJha^{-1}$ mm^{-1}) and I is the rainfall intensity in (mm/h).

4.5 Rainfall distribution uniformity

In simulated rainfall on a plot, uniformity is one the most important measure of determining how spatially distributed the rainfall is on a plot to avoid ponding and over saturation on one side [3]. It therefore measures the equal catches of simulation of rainfall [28]. There are factors that sometimes affect uniformity: this includes; wind, slope and altitude [1]. The degree of uniformity is dependent of the rainfall type. It is estimated using the Christiansen uniformity coefficient (C_u) equation as presented in Eq. (3) [3]

$$C_u = 1 - \frac{SD}{I_m} \tag{3}$$

where C_u is the Christiansen uniformity coefficient; *SD* is the standard deviation of simulated rain over the plot; I_m is the mean simulated rain intensity.

Eq. (2) can further be expressed as in Eq. (4)

$$CU = 1 - \left[\frac{\sum / X_i - X_{m/}}{nX_m} \right] x100 \tag{4}$$

where X_i is the individual rain gauge, X_m is the mean gauge of the rainfall and n is total number of rain gauges.

Spray patterns of different types are obtained from different nozzles. In rainfall simulators, there are two different types of nozzles that are often used based on their mould. Namely; flat and cone spray nozzles. From each of these nozzles there tends to be decrease in uniformity from centre to outward of the sprayed plot [3, 24]. The challenge of rainfall uniformity reducing from centre to outward of the plot can be mitigated by using network of nozzles, taking into consideration the wetted perimeter of individual nozzles. The wetted perimeter depends on the distance of the simulator from the plot for nozzle that produces cone spray, operating pressure [36] in drop forming simulators (DFs) whereas in pressurised simulator (PN) C_u is increased based on increase in pressure and intensity [3]. Many researches have been carried out to estimate uniform application of depths as was used by Christensen to investigate the factors affecting water distribution from group of sprinklers [28, 29, 37–39], but this has been recently criticised based on the fact that is less significant and that size of rain gauge for uniformity and intensity affects the results [14], yet it often used as guide in rainfall simulation. Uniformity of can be more than 90% [31] contrary to sprinkler uniformity standard bench mark of 85% [10].

The methods employed to measure coefficient of uniformity plays a significant role in achieving correct simulated rainfall data [3]. It is therefore difficult to compare the uniformity results of simulated rainfall from different report [31]. In a review, [5] pointed out that drop forming simulators produces higher rainfall uniformity than pressurised nozzle simulator at lower rain intensity. Generally speaking without considering rainfall simulator type, investigator achieved average rain uniformity of 83% within the intensity range of 10 mm/h and 182 mm/h [1, 3, 31].

5. Design requirements of rainfall simulator

To successfully achieve afore listed natural rainfall characteristics, a designer of a rainfall simulator should take into considerations the following phonotypical features; pump pressure, simulators height, plot size and nozzle spacing. Each these physical features have impact on the purpose for which the rainfall simulator is designed.

5.1 Pressure

In pressurised nozzle simulator the choice of pressure is a determining factor to mimic the natural rainfall to the nearest possible outcome [40]. The basis for selecting pressure should be such that balance is stroked among rain intensity, uniformity, rain drop size and kinetic energy, but different researchers are embedded with different approach toward pressure [3]. For example, in an investigation carried out by Cerda et al. indicated that uniformity was obtained at pressure 152 kPa using HARDI-1553-10 single nozzle and anything above this settings resulted to higher rain concentration at the plot boarder and below resulted to concentration of rain at the centre of the plot. The researcher therefore noted that increase in pressure has a maximum limit when targeting at rain uniformity above which decreases the uniformity [3]. In similar research by Sousa-Junior & Siqueira [31], similar trend of results were observed. Simulator under rainfall intensity of 3.1 mm/min, produced uniformity coefficient of 85% at 40 kPa [36]. Comparing the result of [35] with [41] investigation of rainfall intensity at 20 kPa and achieving 1.42–1.58 mm/min with an average rain uniformity of 60%, therefore, the effect of pressure cannot be over emphasised.

In Aksoy et al. [33] investigation, the orifice size was appreciated on examining the effects of pressure on 4-Veejet 8030, 4-Veejet 8050, 5-Veejet 8060 and 5-Veejet 8070 nozzles of different orifices, except for 5-Veejet 8060 nozzle which gave rain uniformity of 83.6% at 33 kPa pressure otherwise the others mimicked uniformity of 82.1, 86, and 88.8% at 40, 42 and 48 kPa respectively. Larger orifice resulted to increase in uniformity though with increase in pressure. According to [14], study on development and calibration of pressurised nozzle simulator observed that uniformity and intensity of modelled rainfall are affected by nozzle pressure disc angular velocity and angle of aperture.

5.2 Nozzle spacing

Nozzle spacing in rainfall simulation is a very vital parameter to be considered in the study of the rain uniformity. Where there is overlapping during spray from two or more nozzles results to higher intensity and uniformity. But report discussion on this has always been mute in literatures [3]. An average CU of 80% was obtained with the use of 4 fixed Veejet nozzles spaced between 2 and 4 m, but when the spacing was reduced to 1, 2 m greater uniformity >86% was achieved [42]. Gabric et al. [34] design Veejet 80,100 nozzle and spaced 100 cm apart to assess intensity and uniformity of simulated rainfall, he achieved a uniformity of 86% at pressures of 40 kPa. Aksoy et al. [32] also studied rain uniformity using a similar nozzle Veejet 8030 and varied nozzle space between 1.45 and 1.25 m at 40 kPa and they achieved CU of 82.1%. A similar trend of results was obtained by [31] using 2-FullJet1/2 SSHH40 nozzle with 1.06 m spacing and varied pressure between 50 and 170 kPa. This shows that the smaller the nozzle spacing, the less pressure required and the larger the spacing the more will be required to mimic good rain uniformity.

5.3 Plot size

The size of a plot is very crucial in the simulation of rainfall most especially in the determination of uniformity. The plot is therefore the predefined seclusion upon which parameter are examined for the purpose of research using simulated rain. It determines the size of the rainfall simulator [3]. Previous research showed that plot area varied from 0.24 [38] to 99 m^2 [43]. Many investigators' results showed that the smaller the plot size for rainfall simulation the higher the uniformity [3]. An example is the result obtained by Sanguesa et al. as cited by [3] with one nozzle used on 1 m × 1 m and 2 m × 2 m plot size they achieved a CU of 91 and 86% respectively. The results gotten when four nozzles arranged in strength line on a plot size of approximately 4.0625 m^2 was 90% [3]. To explicate more on the effect of plot size on uniformity, 4flood jet nozzle was used on two different plot sizes of 3.56 m^2 [10] and 8.84 m^2 [33] and they obtained a corresponding uniformity coefficient of >90% and an average of 85.1%. The aforementioned result confirms that the plot size of a rainfall simulator affects the rain uniformity thus; increase in rainfall simulator plot size will decrease the uniformity. Sometimes the size of plot for rainfall simulation depends on the purpose for which the simulator is designed for. For example [38] selected plot size larger than the simulator top while [5] in a review pointed out some researchers makes use of smaller to obtain good uniformity. In nutshell, the factor determines selection of plot size in rainfall simulator is size of the simulator and the parameter under investigation [23].

Based on the simulator type, drop forming simulators are generally small in area (0.98 ± 68 m^2) which can cover plot size of area 1.07 ± 0.12 m^2 while in the case of pressurised nozzle type of simulator except for those using single; it can be as large as 5.12 ± 1.58 m^2 [3]. Larger plot size in pressurised nozzle requires high pressure at higher height to attain good rain uniformity on the plot. For example, with plot size of 2.8 m^2, rainfall intensity of 1.43–1.58 mm/h and rain uniformity of only 60% was achieved with operating pressure of 20 kPa [41]. These results were not encouraging at all but when pressure of 41 kPa was used on similar plot size of 2 m × 1.5 m (3 m^2) a rainfall uniformity of 95% was achieved as cited by [5].

5.4 Simulator height and kinetic energy

Kinetic energy of simulated rain is being influenced by two major factors; height of simulator and surface of plot, most especially in drop former (DF) simulators

Type/class of rainfall simulator	Drop size (mm)	Area (m^2)	Terminal velocity (m/s)	Uniformity coefficient (%)	Rainfall intensity (mm/h)	Height (m)	Ref.
Drip former RS	4.5	1	0.08	75	0.25–160 (Avg = 80.13)	14	Regmi and Thomson [47]
Drip former RS	3–6 (Avg = 4.5)	0.36	0.023	50–70 (Avg = 60)	78	10–13 (Avg. = 11.5)	Law [11]
Average	**4.5**		**0.05**	**67.5**	**79.06**	**12.75**	
Pressurised RS	0.9–2	0.95	0.63–0.86 (Avg = 0.75)	81.2–88.5 (Avg = 84.85)	55–88 (Avg = 71.5)	1.5–2.5 (Avg = 2)	Ngasoh [48]
Pressurised	2.35–2.55	1.2	1.01	85.7–87.5 (Avg = 86.6)	50.8–152.4 (Avg = 101.6)	4.27	Rick et al. [49]
Average	**2.45**		**0.88**	**85.72**	**86.55**	**3.14**	
Hybrid RS	2.2–8 (Avg = 5.1)	1	0.35	96.5–98.7 (Avg = 97.6)	65–70 (Avg. = 67.2)	2.35	Carvalho et al. [16]
Hybrid RS	2.3	2.5–8	0.123	93	15–120 (Avg. = 67.5)	3.4	Bowyer-Bower and Bur [4]
Average	**3.7**		**0.23**	**95.3**	**67.35**	**2.88**	
Natural rain	0.125–1 (Avg = 0.56)		0.85–5	≥85	15–160 (Avg = 87.5)		Liu et al., [50]

Table 1.
Results of test from different rainfall simulator compared to natural rainfall.

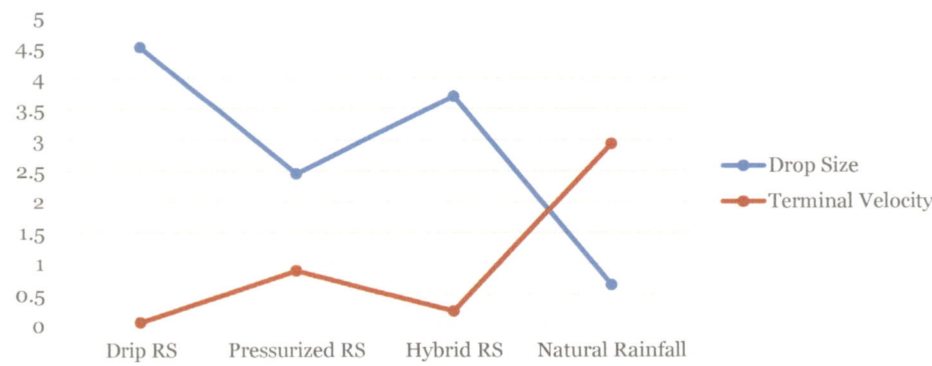

Figure 7.
Representation of rainfall drop size and terminal velocity of different rainfall simulator compared to natural rainfall.

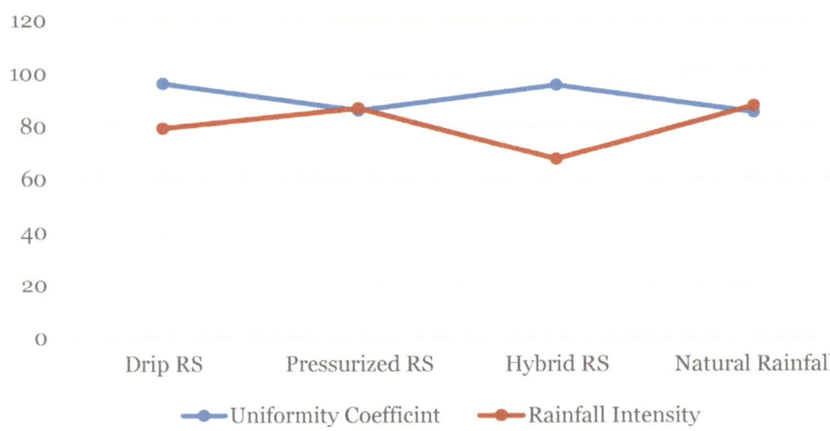

Figure 8.
Representation of rainfall intensity and uniformity coefficient of different rainfall simulator compared to natural rainfall.

[3] requires huge range of height from 7 m [6] and 10 m [44] to reach the terminal velocity. In similar research [45] developed a laboratory rainfall DF simulator, they would achieve the desired kinetic, the dripper was placed at 14 m above the plot. Examining the above results shows that the height of a simulator has significant influence on terminal velocity and kinetic energy. For example low kinetic energy of 5.8 Jm^{-2} mm^{-1} was achieved in a research due to low height of 2 m was used for their simulator [46]. This was also confirmed by when [34] used portable rainfall simulator to control rainfall, some of the rainfall parameters like KE was mimicked at 5 m above the plot to achieve the KE similar to natural rain.

One of the underlined differences between drop former (DF) simulator and pressurised nozzle (PN) is height of the simulator. The pressurised due to the pressure achieves kinetic energy (25 Jm^{-2} mm^{-1}) and D_{50} of 2.19 mm at the height of as low as 2.4 m above the plot as indicated by [24, 33]. According to [5] comparing the results of drop former simulator and pressurised nozzle both positioned at downward spray, pressurised nozzles overestimated the kinetic energy while drip former underestimated the kinetic energy.

After close analysis of the relationships of rainfall simulator components interdependently, [5] further observed that increased in pressure increases the intensity, rain uniformity and kinetic energy. Differences in plot size do not relate any other parameter apart from uniformity. Nozzle spray angle of aperture impacts the nozzle spacing. The research further recommended that any rain simulator designer should take into consideration intensity, kinetic energy and uniformity when designing a rain simulator. **Table 1** showed the results gotten by different researchers using different types of the rainfall simulators.

The average results from the various test indicated that Drop former produces higher rainfall drop sizes followed by hybrid while with pressurised rainfall simulator, an average of 2.5 mm rainfall drop size is produced. That is, among the different types of rainfall simulators, the pressurised rainfall simulator produces small varieties of drop sizes close to that of the natural rain. However, on the terminal velocity, the natural rainfall attains it before reaching it is fall from an infinity distance compared to the on obtainable from simulated rainfall (see **Figure 7**).

Figure 8 compares the uniformity and rainfall intensity of different types of rainfall simulators to the one obtainable from natural rainfall. The findings indicate that Drop former and hybrid rainfall simulator produces higher uniformity coefficient compared to what is obtainable from natural rainfall. While, intensity of a rainfall from pressurised rainfall simulator is similar to the ones obtainable from natural rainfall.

6. Rainfall simulation on non-erodible and erodible surface

For a rainfall simulator to be used to study either on erodible or non-erodible surface, it needs to achieve rainfall characteristics close to those of natural rainfall, it needs to be portable and easy to control [3].

Furthermost of the research on erodible surface have involved erosion, infiltration and tillage studies [24, 51]. In disparity, the process concerning urban wet weather studies involved non-erodible surface and were defined based on pollutant volume and the corresponding discharge volume [3, 52]. In run off and sediment yield studied by [53] from an erodible watershed and non-erodible watershed using 10 modelled precipitation event, they achieved a runoff volume and sediment load of 5.5 ± 2.7 and 5.5 ± 2.3 respectively, and the proportion of precipitation to runoff volume was on the average 14.5%. The simulated result was greater than when it was done on non-erodible soil surface. A conclusion was also made by [51, 54] that drop size and the fall velocity are given basic attention in the study of erosion and infiltration model involving erodible surfaces and [53] also noted that simulation on non-erodible surface increased runoff volumes linearly and peak flow rate exponentially and served as means of control of sediment load and flow rate by its spatial characteristics.

7. Conclusion

First of all, the method employed to accumulating runoff on non-erodible and erodible surface not the same. Simulating precipitation and collection of runoff on non-erodible surface is more challenging because non-erodible surface are mostly tiled surfaces where excavation is controlled. Recovering of the runoff from non-erodible surface is the priority of researchers but the task is difficult. To overcome the difficulties in regenerating the runoff from an urban non-erodible surface.

Secondly, take note of the length and slope of the study area in the study of erosion and infiltration as they are important requirement in simulation.

Thirdly, pressurised nozzle simulator will be suitable for simulating reasonable intensity, runoff and rain depth most especially for nonpoint source study on non-erodible surfaces because the controlling intensity will be limited using drop former simulator.

Fourthly, in the simulation of drop size and distribution, water quality should be taken note of. Though it may not be significant in the simulation of infiltration and soil erosion, but in urban water quality simulation which deals with measurement of pollutant level it is a very important factor to consider. In an investigation carried out in Malaysia [55], water quality presented a challenge in simulating intensity drop size, drop size distribution and uniformity using drop former simulator. As water is stored and kept for long period of time algae and other micro-organism may develop in it or around the dripper. This challenge is predominant in drop forming simulators and less in pressurised nozzle simulator because the pressure applied at the nozzle orifice reduces the risk of clogging. To minimise the challenge of clogging of dripper and nozzle orifices, screen should be provided at suction point or water source.

Duration of experiment on non-erodible surface using rain simulator is an important requirement. To overcome the delay in runoff generation on studying runoff on non-erodible surface which is predominant in drop former simulator, pressurised nozzles are preferable because it offers reasonable amount of runoff with short time. In contrast, on erodible surface, drop former simulators are preferred especially in the study of infiltration.

Rainfall uniformity is achieved higher in drop forming and hybrid simulators which is a good requirement for erodible surface that include infiltration studies where the interest is on measuring downward filtered water on the plot. Simulating on erodible surface, saturation of the plot surface is slower than simulation on non-erodible surface. On the non-erodible surface the study interest is runoff collection. The researcher further recommended that mounting and dismounting of rainfall simulator should be flexible.

Finally, to achieve a good rainfall distribution uniformity using rainfall simulator, the plot must be smaller than the wetted perimeter of the simulator most especially for outdoor simulator. In the case of indoor rainfall simulator, the plot can be larger than the wetted perimeter but consideration can only be given to collectors around the wetted perimeter.

Author details

Felix Gemlack Ngasoh[1]*, Constantine Crown Mbajiorgu[2], Matthew Boniface Kamai[1] and Gideon Onyekachi Okoro[2]

1 Department of Agricultural and Bioresources Engineering, Taraba State University, Jalingo, Nigeria

2 Department of Agricultural and Bioresources Engineering, University of Nigeria, Nsukka, Nigeria

*Address all correspondence to: felixngasoh@gmail.com

References

[1] Grismer M. Standards vary in studies using rainfall simulators to evaluate erosion. California Agriculture. 2012;**66**:102-107

[2] Sobrinho TA, Macpherson GH, Gómez JA. A portable integrated rainfall and overland flow simulator. Soil Use and Management. 2008;**24**:163-170

[3] Yakubu ML, Yusop Z. Adaptabilty of rainfall simulators as a research tool on urban sealed surfaces. Hydrological Sciences Journal. 2017;**62**:996-1012

[4] Bowyer-Bower TAS, Burt TP. Rainfall simulators for investigating soil response to rainfall. Soil Technology. 1989;**2**:1-16

[5] Morin J, Goldberg D, Seginer I. A rainfall simulator with a rotating disk. Transactions of ASAE. 1967;**10**:74-0077

[6] Wildhaber YS, Bänninger D, Burri K. Evaluation and application of a portable rainfall simulator on subalpine grassland. Catena. 2012;**91**:56-62

[7] Lassu T, Seeger M, Peters P. The Wageningen rainfall simulator: Set-up and calibration of an indoor nozzle-type rainfall simulator for soil erosion studies. Land Degradation and Development. 2015;**26**:604-612

[8] Palmer RS. The influence of a thin water layer on waterdrop impact forces. International Association of Hydrological Sciences. 1963;**65**:141-148

[9] Mutchler C, Moldenhauer W. Applicator for laboratory rainfall simulator. Transactions of ASAE. 1963;**6**:220-0222

[10] Blanquies J, Scharff M, Hallock B. The Design and Construction of Rainfall Simulator, presented at the International Erosion Control Association (IECA), 34th Annual Conference and Exposition. Nevada: Las Vegas; 2003

[11] Laws JO. Measurements of the fall velocity of water drops and raindrops. Transactions of the American Geophysical Union. 1941;**22**:709-721

[12] Meyer LD, McCune DL. Rainfall simulator for runoff plots. Agricultural Engineering. 1958;**39**:644-648

[13] Wilson TG, Cortis C, Montaldo N. Development and testing of a large, transportable rainfall simulator for plot-scale runoff and parameter estimation. Hydrology and Earth System Sciences. 2014;**18**:4169-4183

[14] Pall R, Dickinson WT, Beals D. Development and calibration of rainfall simulator. Canadian Agricultural Engineering. 1983;**25**:181-187

[15] Egodawatta P. Translation of small-plot pollutant build-up and wash-off measurement to urban catchment scale [PhD]. Queensland University of Technology; 2007

[16] Carvalho SCP, de Lima JLMP, de Lima MIP. Using meshes to change the characteristics of simulated rainfall produced by spray nozzles. The International Soil and Water Conservation Research. 2014;**2**:67-78

[17] Darboux F, Renaux B, Cottenot L. The indoor rainfall simulator of INRA-Orleans (France). In: Rainfall Simulator Workshop. Germany: Trier; 2011. p. 13

[18] Loch R, Robotham B, Zeller L. A multi-purpose rainfall simulator for field infiltration and erosion studies. Soil Research. 2001;**39**:599-610

[19] Nadal-Romero E, Lasanta T, Regüés D. Hydrological response and sediment production under different land cover in abandoned

farmland fields in a mediterranean mountain environment. Boletín de la Asociación de Geógrafos Españoles. 2011;**55**:303-323

[20] Wang J, Hong Y, Li L. The coupled routing and excess storage (CREST) distributed hydrological model. Hydrological Sciences Journal. 2011;**56**:84-98

[21] Wang N, Jiao JY, Lei D. Effect of rainfall erosion: Seedling damage and establishment problems. Land Degradation & Development. 2014;**25**:565-572

[22] Ziadat FM, Taimeh A. Effect of rainfall intensity, slope, land use and antecedent soil moisture on soil erosion in an arid environment. Land Degradation & Development. 2013;**24**:582-590

[23] Battany M, Grismer M. Development of a portable field rainfall simulator for use in hillside vineyard runoff and erosion studies. Hydrological Processes. 2000;**14**:1119-1129

[24] Abudi I, Carmi G, Berliner P. Rainfall simulator for field runoff studies. Journal of Hydrology. 2012:76-81

[25] Gabriel D. Use of field and laboratory rainfall simulators for assessing the factors of the soil erosion process. In: Proceeding of Rainfall Simulator Workshop. Deutsche; 2011. p. 9

[26] Kathiravelu G, Lucke T, Nichols P. Raindrop measurement Tchniques. Water. 2016;**8**(29):1-20

[27] Herngren LF. Build-up and Wash-off Process Kinetics of PAHs and Heavy Metals on Paved Surfaces Using Simulated Rainfall [Ph.D.]. Brisbane, Australia: Russian Academy of Science, Queensland University of Technology; 2005

[28] Nassif SH. Hydrological Studies of a Laboratory Catchment Area under Rainfall Simulation. Manchester: University of Salford; 1973

[29] Lora M, Camporese M, Salandin P. Design and performance of a nozzle-type rainfall simulator for landslide triggering experiments. Catena. 2016;**140**:77-89

[30] Nash JE. Determining runoff from rainfall. Proceedings of the Institution of Mechanical Engineers. 1958;**10**:163-184

[31] Sousa-Júnior SF, Siqueira EQ. Development and Calibration of a Rainfall Simulator for Urban Hydrology Research, Presented at the 12th International Conference on Urban Drainage. Proto Alegre/Brazil; 2011

[32] Horne MA. Design and Construction of a Rainfall Simulator for Large-Scale Testing of Erosion Control Practices and Products [Master Research]. Alabama: Civil Engineering, Auburn University; 2017

[33] Aksoy H, Unal NE, Cokgor S. A Rainfall Simulator for Laboratory-Scale Assessment of Rainfall-Runoff-Sediment Transport Processes over a Two-Dimensional Flume. CETENA: Elsevier; 2012. pp. 63-72

[34] Hignett CT, Gusli S, Cass A, Besz W. An Automated Laboratory Rainfall Simulation System with Controlled Rainfall Intensity, Raindrop Energy and Soil Drainage. Vol. 8. Soil Technology, Elsevier; 1995. pp. 31-42

[35] Foster GR, McCool DK, Renard KG. Conversion of the universal soil loss equation to SI metric units. Journal of Soil and Water Conservation. 1981;**36**:355-359

[36] Gabric O, Prodanovic D, Plavsic J. Uncertainty assessment of rainfall simulator uniformity coefficient:

40th anniversary of faculty of civil engineering. Journal of the Faculty of Civil Engineering. 2014:661-667

[37] Christiansen JE. Irrigation by sprinkling. In: College of Agriculture, Agricultural Experiment Station. Vol. 670. Berkeley. Califonia: University of California; 1942

[38] Cerdà A, Ibáñez S, Calvo A. Design and operation of a small and portable rainfall simulator for rugged terrain. Soil Technology. 1997;**11**:163-170

[39] Holtan HN, Kirkpatrick MHJ. Rainfall, infiltration and hydraulics of flow in runoff computations. Transactions of the American Geophysical Union. 1950;**31**:771-779

[40] Tossell RW, Dickinson WT, Rudra RP. A Portable rainfall simulator. Canadian Agricultural Engineering. 1987;**29**:155-162

[41] Fister W, Iserloh T, Ries JB. A portable wind and rainfall simulator for in situ soil erosion measurements. CATENA. 2012;**91**:72-84

[42] Bauer M, Kavka P, Strouhal L, Dostál T, Krása J. Rainfall Simulators - Innovations Seeking Rainfall Uniformity and Automatic Flow Rate Measurements. Czechia: University in Prague; 2016

[43] Moore GW, Jones JA, Bond J. How soil moisture mediates the influence of transpiration on streamflow at hourly to interannual scales in a forested catchment. Hydrological Processes. 2011;**25**:3701-3710

[44] Morin J, Cluff CB, Powers WR. Realistic Rainfall Simulation for Field Investigation, Presented at the 55th Annaual Meeting of the American Geophysical Union Washington D.C; 1970

[45] Regmi TP, Thompson AL. Rainfall simulator for laboratory studies. American Society of Agricultural and Biological Engineers. 2001;**16**:641-647

[46] Iserloh T, Ries JB, Arnáez J. European small portable rainfall simulators: A comparison of rainfall characteristics. Catena. 2013:1-13

[47] Regmi TP, Thompson AL. Rainfall simulators for laboratory studies. Applied Engineering in Agriculture. 2000;**16**:641-647

[48] Ngasoh FG. Design Construction and testing of a Hydrological cycle demonstration system [Master Research]. Nsukka: Agricultural and Bioresources Engineering, University of Nigeria; 2018

[49] Ricks MD, Horne MA, Faulkner B. Design of a pressurized rainfall simulator for evaluating performance of erosion control practices. Watermark. 2019;**11**(23):1-18

[50] Liu N, Lin J, Che S. Raindrop characteristics analysis (Oct 25, 2015) of natural rainfall in Zhengzhou city of Yello River basin. In: IOP Conf.: Earth and Environmental Science. Vol. 128. 2018

[51] Parsakhoo A, Lotfalian M, Kavian A. Calibration of a portable single nozzle rainfall simulator for soil erodibility study in hyrcanian forests. African Journal of Agricultural Research. 2012;**7**:3957-3963

[52] Yu VB. Runoff generation and storage in watershed. Hydrologic Cycle. 2011;**III**:10

[53] Corbett CW, Wahl M, Porter DE. Nonpoint source runoff modeling a comparison of a forested watershed and an urban watershed on the South Carolina coast. Journal of Experimental Marine Biology and Ecology. 1997;**213**:133-149

[54] Hall MJ, Johnston PM, Wheater HS. Evaluation of overland flow models using laboratory catchment data I. an apparatus for laboratory catchment studies. Hydrological Sciences Journal. 1989;**34**:277-288

[55] Yakubu ML. Modelling of Road surface pollution buildup and washoff using rainfall simulator [PhD]. Malaysia: Civil Engineering, Universiti Teknologi Malaysia; 2015

Chapter 8

The Role of Biosensor in Climate Smart Organic Agriculture toward Agricultural and Environmental Sustainability

Kingsley Eghonghon Ukhurebor

Abstract

The application of climate smart agriculture technique has been identified as an important aspect in proffering solutions for most of the challenges to climate change mitigation as well as environmental and agricultural sustainability. Several biosensors and biosensing machineries such as nanoparticles/nanomaterials, polymers and microbes built-biosensors as well as their applications are now being used in most part of the world for solving some of the challenges in agricultural activities, food production and its sustainability. However, it is significant to assimilate multifaceted methods for sustainable development of more effective biosensors that can possibly be used for diverse applications especially in the area of climate smart organic/biological agriculture for environmental sustainability. Smart monitoring employing biosensors will ensure that biochemical and other categories of contaminants are kept at bay from conceding the quality and safety of food via the mitigation of pest and pathogens that could affect agricultural produce. Hence, this study will attempt to provide an overview of what has been done from previous studies on biosensing technologies and their wide application in climate smart organic/biological agriculture as well as their role in environmental sustainability, and this will assist in proffering useful suggestions for future research studies as future contribution to knowledge for the advancement in agricultural and environmental sustainability.

Keywords: agriculture, biosensors, environment, food, sustainability

1. Introduction

Agriculture is an imperative and steady aspect of human existence, owing to fact that human survival depends on agricultural produce. Our engagement in agricultural activities have potentially and continually been of great assistance in the production and availability of food, raw materials, chemical and several other industrial resources [1, 2]. The agricultural sector as reported by "the Food and Agriculture Organization of the United Nations (FAO)" is faced with several problems such as failures in the market system and barriers in the trade system, uneven and futile socio-economic strategies, insufficient information, availability of finance and infrastructures, pressure at a result of upsurge in the population and insufficiency resources, agronomic practices, unsustainability and dilapidation in the environment, etc. [3]. These problems are further confronted by the influences of climate inconsistency and deviations as agriculture primarily dependent on climate variables/parameters [4–6]. Consequently, Akrofi-Atitianti et al. [5] in their study reported that agricultural sector in the developing nations (like Africa and most other developing regions) remains one of the furthermost susceptible sectors to these problems of climate inconsistency and deviation. The issue of food safety and security as well as deficiency in food supply and climate change have a very strong relationship and according to Karimi et al. [2], it will be appropriate to always consider them to together. In light of this, "the United Nations Framework Convention on Climate Change (UNFCCC)" and "the Intergovernmental Panel on Climate Change (IPCC)" have always emphases importance of agriculture and have incessantly placed great priority on agricultural activities [3, 4].

According to Karimi et al. [2], the influences of climate change on agricultural activities are still lagged with some uncertainties. Nevertheless, climate modification is anticipated to unpleasantly affect agricultural sector as well as other sectors and human activities globally; this would be as a result of the vicissitudes in precipitation, temperature, carbon dioxide pollination and other weather parameters/variables [7–10]. Consequently, climate adaptation techniques are ultimately essential for mitigating these increasing climate/weather actions in our environment [6, 11–14]. According to Abegunde et al. [15], climate smart agriculture is a substantial aspect in proffering solutions for both climate change mitigation, agriculture and environmental sustainability. They reported that agricultural activities can contribute significantly to climate change mitigation in the following ways:

a. The avoidance of further deforestation and conversion or/and alteration of wetlands (marshlands or swamplands) and grasslands (savannahs).

b. The intensification and spiraling in the storage of carbon in vegetation and soil.

c. The reduction of current and the avoidance of future upsurges in greenhouse gases (GHGs) emissions from nitrous oxide, methane and other forms of GHGs.

Efforts genuinely gear toward the reduction of GHG emissions should be embrace agriculture. According to Fanen and Olalekan [16], some of the most important agricultural products are possible of filling the gaps between recent produces and the produces that have the potential for improving inputs and management as well as the promotion of truncated GHS emission possibilities. Climate smart agriculture has some exceptional possibilities in tackling food safety, security, adaptation and moderation tenacities [15, 16]. It has been reported that climate smart agriculture is a dependable alternative that can assist in undertaking the food insecurity issues that are alleged to be caused by the altering of the climate/weather [15, 16]. However, some developing countries have realized that some of these concepts of climate smart agriculture that have been recommended as solutions to existing problems are somehow not too suitable in their contexts as a result of some environmental capriciousness [2, 15, 16]. Besides, agriculture played a fundamental part in the alleviation of poverty and to enact major undesirable influences that climate change is expected to have on several regions globally.

Supposedly, early accomplishment in climate smart agriculture has been recognized as an indispensable means of capacity building as well as skill and guide for future opportunities [15]. However, it is desirable to have a proper meaning of what is meant by smart system before exploring to climate smart agriculture practice. Hence, according to Abegunde et al. [15] "a smart system or product is that which facilitates the interface of a system with persons/users and is able to acclimate the framework of the user without compelling the user to acclimate to it". Smart system may comprise of the following characteristics [2, 15]:

a. Capability to collaborate with other devices.

b. Adaptability to acquire and improve the compatibility between its functioning and its environment.

c. Self-sufficiency, which indicates that the system can function without intrusion from the user.

d. Capability to network with person via natural interface.

e. Multi-purposeful which indicates a single product is capable of executing multiple roles.

f. Personality which indicates the system is proficient to be active and accomplish the features of credible personality.

g. Reactivity, which indicates that the system can respond to its environment in a special way.

A smart system is capable to carry out an integral approach, from sensing to acting, to carry out optimal on-line control for performance or product quality through smart sensing techniques, besides the use of biosensors has contributed to the advancement of climate smart agriculture. Biosensors technology has the potential to improve agricultural productivity as well as food, chemical and other industrial innovative tools and techniques for the monitoring and management of swift infection disease diagnostic, the capacity enhancement of plants for the absorption of nutrients, the capacity enhancement of animal production, etc. [16]. In other to address the contribution of agricultural activities to these problems of alterations in the climate system, climate smart agriculture is gradually being indorsed in most parts of the world to assist in the integration of the economy, social and environmental extents of sustainable development in building on the three core aspects, viz.: "sustainably increasing agricultural productivity and revenues; acclimating and building resilience to climate change and; reducing and/or removing GHGs emissions relative to conventional practices" [3]. It is to be reported that climate smart agriculture has not yet be fully adopted in most developing countries (Africa inclusive); this is attributed to the limited understanding of the constraints of these countries to effectively implement the adaptation approaches faced by those involved in agricultural activities across these regions [5, 17]. Even though most developed nations of the world are beginning to adopt and apply climate smart agriculture, there is still a great deal to be done for its improvement. In light of this, biosensor in climate smart organic agriculture need to be incorporated, and this will definitely play a significant role in agricultural and environmental sustainability.

Due to the incessant growing of the world's population which according to the United Nations (UN) [18] is projected to reach around nine billion by the year 2050 from the present estimated eight billion is considered a time bomb due to the fact that upsurge in population will obviously translate to equivalent increase in food demand. Smart monitoring employing biosensors will ensure that biochemical and other categories of contaminants are kept at bay from conceding the quality and safety of food as well as the pest and pathogens that could affect agricultural produce. Biosensors are also deployed for the purpose of measuring alcohol, carbohydrates, acids, etc. Hence, this chapter will attempt to present an assessment of what has been done from previous studies in biosensing technologies for climate smart organic/biological agriculture as well as their role in agricultural and environmental sustainability vis-à-vis food safe/security and climate change that are been explored by researchers in the area of biosensors technology for the improvement of agriculture. The limitations faced with some of the prominent techniques especially as it relates to climate smart organic/biological agriculture will be highlighted; this will evidently assist in proffering useful suggestions for future research studies as future contribution to knowledge for the advancement in agricultural and environmental sustainability.

2. Application of biosensors in climate smart organic agriculture

Biosensors are diagnostic devices that combine biological constituents and transducers for the discovery of sample like metabolites, drugs, microbial load, contaminants, control parameters, etc. They do so by translating biochemical reactions into quantifiable physiochemical signals such as electrical signal which in turn measure the amount of sample that are used for the discovery of analyte concentration [19–22]. Biosensors have several applications in the diverse fields or areas such as medicinal discovery and diagnosis, food protection and processing, defense and security, environmental management, etc. [21–23]. There are several types of biosensors used in the environment as it relates to soil, water and air in the area of in climate smart biological/organic agriculture and these biosensors depend on the sensing rudiments or transducers. Ever since the first discovering of the glucose biosensor in 1956, by Prof. L.C Clark Jnr, which however came in limelight commercially in 1975 [24], there are now several biosensors discovered for various commercial purposes. These contemporary biosensors have wider range of applications which offers additional specific, sensitive, fast, tangible and multiplicative results compared to previous chemical sensors [19, 20, 23].

Presently, with the advancement of nanotechnology, innovative nanomaterials are now being invented and their innovative features as well as their applications in biosensors [23, 25]. Nanomaterials-built biosensors, encompass the combination of biotechnology, molecular engineering, chemistry, physics, environmental and material science. These various fields have been of great assistance in advancing the understanding and specificity of biomolecule discovery, the ability of detecting or manipulating atoms and molecules, biomolecular recognition, pathogenic diagnosis as well as the monitoring and management of agriculture and the environment in general [23, 25]. The application of various biosensors such as nanoparticles/nanomaterials, polymers and microbes built-biosensors for agricultural and environmental activities have assisted in the reduction of the quantity of chemicals spread, reduction in nutrient losses in fertilization and upsurge in the yields via the reduction of pests and diseases for the enhancement of nutrients [26].

Biosensors are broadly categorized into two classes which are based on sensing components and transduction modes. The sensing components consist of enzymes, antibodies (immunosensors), micro-organisms (cell biosensors), biological tissues and organelles. While, the transduction modes hinge on the physiochemical variation resulting from sensing components. Accordingly, dissimilar transducer biosensors can be piezoelectric, electrochemical, calorimetric and optical [19, 27]. As reported by Reyes De Corcuera and Cavalieri [27], the common types of piezoelectric transducer biosensors are acoustic and ultrasonic; the common types of electrochemical transducer biosensors are amperometric, conductometric and potentiometric; while the common types of optical transducer biosensors are absorbance, fluorescence and chemiluminense. According to Arora [20], biosensors can also be categorized based on the period/order they were discovered. In these categories we have first-generation biosensors, second-generation biosensors and third-generation biosensors.

The first-generation biosensors: These biosensors are the modest approach involving the unswerving discovery of either increase of an enzymatically produced product or decrease of a substrate of a redox enzymes using natural mediator for electron transfer. Examples are glucose biosensor which uses enzyme glucose oxidase and oxygen detecting decrease in oxygen level or increase in hydrogen peroxide corresponding to the level of glucose.

The second-generation biosensors: They the biosensors that use non-natural redox mediators like ferricynide, quinones and ferrocene for the movement of electron which increases the reproducibility and sensitivity. Examples are self-monitoring amperometric glucose biosensors.

The third-generation biosensors: These are biosensors wherein the redox enzymes which are immobilized on the electrode surface in such a manner that direct electron transfer is possible between the enzyme and transducer. According to Borgmann et al. [19], it uses organic conducting material like "Tetrathiafulvalnetetracynoquinodi Methane (TTF-TCQN)".

As mentioned earlier, with the present advancement of nanotechnology, innovative nanomaterials are now being invented and their innovative features as well as their applications exist in biosensors [25]. However, it was in 1962 that Clark and Lyons invented the first biosensor that measure glucose in biological samples which utilized the strategy of electrochemical detection of oxygen or hydrogen peroxide via controlled glucose oxidase electrode and even since then, incredible improvement has been attained both in the skill involved and the applications of biosensors with advanced tactics involving nanotechnology, electrochemistry and bioelectronics [19, 21, 23, 24, 28–30]. The discovery of biosensors as an influential and pioneering diagnostic device (which has to do with biological sensing component with several applications) has undoubtedly espoused dominant importance in various fields. Its utilization has attained some significant application in the field of pharmacology, biomedicine, environmental science, food protection and processing (agriculture). Biosensors discovery have led to the development of accurate and influential diagnostic tools by means of biological sensing component as biosensor [21, 22]. According to Turner [30], the technical approaches used in biosensors are built on label-built and label-unrestricted detection. Label-built detection is primarily dependent upon the explicit features of label composites to target detection. Nevertheless, these categories of biosensors are reliable; but are habitually involve in the combination of explicit sensing components fabricated with restrained target protein. On the other hand, label-unrestricted technique allows the detection of the target molecules/particles that are not categorized or hard to tag [31, 32]. Topical interdisciplinary approaches of biotechnology and electronics technology paved way for evolving label-unrestricted biosensors for several detection approaches with numerous applications in the areas/fields of medical science and environmental science.

The major distinctive components in a biosensor which are illustrated in **Figure 1** as a block prototypical distinctive biosensor with a processor and display unit according to Mehrotra [33] are:

a. **Detecting/Sensing Component:** This is also known as a biorecognition component; as in the case of a glucose sensor, the biorecognition component is a deactivated glucose-sensitive enzyme.

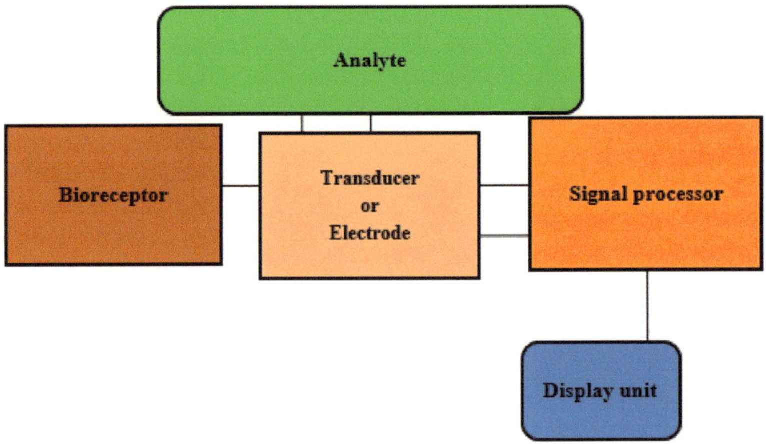

Figure 1.
Block prototypical of a distinctive biosensor.

b. **The Transducer:** The chemical, biochemical, organic, structural or physical device that interpret discrepancies in the target biophysical variables like oxygen, glucose, etc. to a physically quantifiable output signal and/or vice versa.

c. **Signal Processor:** This could be an electrical/electronic device with/without a display system, a processor and an amplifier.

Biosensor machineries are also been applied in agriculture and environmental management/monitoring. According to Verma and Bhardwaj [34], this is another important aspect wherein biosensor technology is beginning to gain grounds. These biosensor machineries in agriculture and environmental monitoring/management will undoubtedly assist in the swift identification of pesticidal deposits in order to avert the corresponding health dangers in form of climate smart organic/biological agriculture [16, 26, 34–38]. According to Verma and Bhardwaj [34], the traditional or conventional means, such as "high-performance liquid chromatography, capillary electrophoresis and mass spectrometry" are efficient for the investigation of environmental pesticides; hitherto, there are some restrictions such as intricacy, time-intense measures, necessity of high-end devices and operative proficiencies. Therefore, even if it is believed that unpretentious biosensors have great advantages; hitherto, it is not easy to invent integrated biosensors that can analyze several categories of pesticides. Hence, steady enzyme-built biosensors have been invented for understanding the physiological (biological and physical) influence of pesticides in the environment, food security and quality management [34, 36, 39]. In the study carried out by Pundir and Chauhan [39], they reported that acetylcholinesterase inhibition-built biosensors have been invented. Over the years, for the purpose of swift analysis, this method has received great improvement with additional topical developments in acetylcholinesterase inhibition-built biosensors including immobilization means as well as other diverse approaches for fabrication [21]. In the same way, piezoelectric biosensors have been established for sensing the organophosphate and carbamate environmental influence of pesticides [36]. Organochlorine pesticides are recognized for affecting the ecosystem where pesticides such as endosulfan cause substantial environmental impairments [40]. Organochlorine pesticides have been reported to cause alteration in the reproductive system of in both male and female fish disparately [40], and in view of these facts, the discovery of biosensors for detecting aquatic ecosystem would have more consequence as a result of biomagnification [21]. In handling this quest, electrochemical biosensors have experienced revolution with swift advances in the fabrication as well as the use of constituents like nanomaterials [21, 24].

At this juncture, it is of great significant to place distinct prominence for the selection or collection of receptors for biosensor advancement, the use of diverse transduction procedures and fast screening approaches for the applications of biosensor in agricultural activities (food production, security and safety) as well as environmental protection, monitoring and management. To aid this, biosensor fabrication appears to be vital and the improvements in this aspect have been absolutely elucidated by several researchers.

3. Biosensor in climate smart organic agriculture

The main challenges faced in agriculture vis-à-vis food safety and sustainability are emphasis on three basic aspects as reported by Neethirajan et al. [38], viz.:

a. Nanomaterials and their application in sustainable agriculture challenges.

b. Energy sustainability challenges.

c. Commercialization of sustainable technology challenges.

Nanotechnology is one of the foremost applications in agricultural monitoring and management. It has several valuable possessions and applicability [38]. According to Prasad et al. [41], it has all it takes to improve food safety and quality, enhance the absorption capability of soil nutrient, increase agriculture inputs, and upsurge the potentials in the miniaturized device measurement. Supposedly, nanotechnology has been used effectively for the following: precision in farming machinery (agricultural precision), smart feed management, food waste management, production of agro-chemicals/agro-materials such as nano-pesticides, nano-herbicides and nano-fertilizers, labelling and packaging of agricultural products, and several other agricultural fields [38]. According to Neethirajan et al. [38], the use of nanotechnology for agriculture as it relates to food sustainability is likely to cause some consequences in the upcoming years. This collaborated the study of Dasgupta et al. [42], that notwithstanding the benefits from the recent combination of nanomaterials/nanoparticles and animated charcoal for the enhancement of antimicrobial possessions, food grade nano-emulsion applied in fruit juice, integrated nano-microbials used as water sterilizers, effective nutraceutical nano-delivery and improved plant extracts conjugated by means of nano-packaging could have some consequences as well. A major emphasis in nanotechnology is in its application for agricultural precision (precision in farming machinery), wherein plant excerpts from its main parts such as leaves, flowers, stems and roots, from various species have been effectively integrated into nanoparticles/nanomaterials [42, 43].

Nanomaterials/nanoparticles have all it takes improve green synthesis in a sole/single-step by means of ion and metal diminishing implications; this according to Prasad et al. [41], is auspicious for the application of room temperature, easy-use, adjustable and climbable as well as eco-system friendly. During green synthesis, co-enzymes and solvable metabolites like phenolic composites, alkaloids as well as terpenoids are wholly condensed to nanoparticles/nanomaterials. Intrinsically, nanoparticles/nanomaterials are known as “magic bullets” resulting in improved plant development, location precise delivery of nutrients and amplified plant infection or disease resistance.

One of the utmost substantial challenges in nanotechnology is in the development of consistent rick-advantage evaluations by means of standardized assessment and procedures. The establishment of reliable and standardized procedures in nanomaterial/nanoparticle measurement, classification and assessment of their effect on living organism and the environment as well as the involvement of all relevant stakeholders such as farmers, agents of food industries, non-governmental organizations etc. in a dialog of public support and consumer acceptance [41, 44]. These challenges in sustainable energy can be effectively taken care of by the application biological or organic solutions. According to Adesina et al. [37], some main applications have been explored in applied organic or biological for the generation of energy:

a. Biofuels could be produced, deposited, transformed and renewed to bio-electricity in order to expressively diminish the cost of producing solar electricity. This can be accomplished by means of leveraging through the intake of H_2 or electron lashing carbon fixing metabolism, to simplify the combination by means of photovoltaics effectiveness in a process known as electro-photosynthesis [42].

b. Hydrogen-built electrosynthesis is one of the furthermost efficacious bioengineering energy creation set-ups. It exhibits exceptional properties such as high effective bioenergy storing capacity for electrical energy of about 80%, lengthy distance transportability with least energy forfeiture, hydrogen oxidation in microorganisms involving "Nicotinamide Adenine Dinucleotide (NAD^+; $C_{21}H_{27}N_7O_{14}P_2$)" decrease diminishing potential discrepancy and affordability as a result of lesser cell-protein necessities of hydrogen oxidation [38].

c. Electron transmission can extracellularly arbitrate electro-synthesis efficiently. This can be done reproducible by means of a nanostructured surface to simplify the of creation bio-film. It could prance the necessity of protracted surface area and improve the transfer of hydrogen electron [38].

d. Applied organic or biological energy creation would be significantly improved by means of the invention of several other machineries. Such innovative machineries are; gene engineering, whole genome engineering, protein engineering, and biosensing [38].

These innovative machineries will curiously enhance the development, production and generation biofuel. Apparently, since one of the furthermost protuberant applications of applied biology is in the area of sustainable energy; hitherto, expectedly biofuel is to become the furthermost positioned procedures in for apprehending and storing solar energy with minimum costs [38]. Presently, the challenges facing the development, production and generation biofuel are: the energy generation effectiveness and scale; competence investigation in cell self-assembly as well as duplication monitoring and management, and antagonistic environmental consequences [38]. Anticipatedly, in the forthcoming years, biofuel is to advance and extend sources of traditional energy to reutilize and replicate the generating constituents of energy and to improve hybrid energy photosynthesis [38]. In **Table 1** outlines some of the aspects where biosensors are deployed in agricultural activities.

However, the commercialization or industrialization of sustainable machineries in the agri-food scope is ongoing via some core emphases such as; biosensor commercialization or industrialization, sensing technology commercialization or industrialization, and intelligent agricultural/food commercialization or industrialization (such as climate smart biological or organic agriculture) [26]. In biosensor

Transduction	Electrode	Analyte sensed	Applications
Electrochemical magnet immunosensing [45]	Magnetic graphite-epoxy composite (m-GEC)	Salmonella in milk	The GEC holds a distinct feature of hybridization that allows the pathogens' DNA to be immobilized instantly. The procedure does not need reagents, and provides swift detection.
Electrochemical magnet immunosensing [46]	m-GEC	b-lactamase resistance in *Staphylococcus aureus*	GEC products have feeble generic adsorption for either DNA samples or enzymes labels. They do not need blocking phases on the transducer's free sites to moderate generic adsorption.
Gold nanoparticle-based [47]	GEC	Salmonella IS200	This is a good substrate for enhanced and directed immobilization of biomolecules with exceptional transductive features for the fabrication of a several of electrochemical biosensors, like immunosensors, genosensors, and enzyme sensors.
Amperometric electrochemical immunoassay [48]	Platinum (Pt) working electrode, Ag/AgCl reference electrode and a Pt counter electrode.	Staphylococcus aureus in food samples such as milk, cheese, and meat	This has been proven to be fast, operative and reproducible, and can be employed to sense specific pathogenic microbes via antibodies against precise antigens.
Multiplexing optical (luminescence) [49]	na	*E. coli* O157: H7, S. typhimurium and Legionella pneumophila	The entire quantification and calibration assay period is 18 min, aiding extremely swift analyses.
High-density microelectrode array [50]	na	*E. coli* O157: H7 bacteria in food materials	It is field-deployable, easy to use, compact, and reagent-less and provides result in minutes compared to conventional procedures.
Flow-type antibody sensor [51]	quartz crystal microbalance chip	E. coli in drinking water, beef, pork, and dumpling	The sensor quantifies changes in frequency as a result mass deposits that are designed by antigen-antibody interface.
Acoustic-based biosensor (the Quartz Crystal Microbalance) [52]	n.a	DNA detection	This enhances the processing of time by circumventing gel electrophoresis and can be combined in a diagnostic laboratory or an automated lab-on-a-chip device for plant pathogen diagnostics as a routine detection device.

Table 1.
Biosensor applications in agricultural activities.

commercialization or industrialization, important aspects for the determination of its commercialization or industrialization are simpler sample pre-treatment, bioreceptor steadiness, multi-detecting/multi-sensing features, impoverishment/miniaturization, quicker testing period, wireless accessibility and affordability [53]. Some the foremost properties of commercialized accessible well-known biosensors industries are their simple structure, reduced sizes and ideal potentials for "point of care" applications [38]. They target food composition, progression monitoring and management as well as food safety and security such as allergens, pathogens, toxins, pollutants/contaminants and additives have been reported that the industries for food quality biosensors purpose is primarily from the following metabolites; "glucose, sucrose, glycerol, cholesterol, creatinine, alcohol, methanol, lactate, lactose, glutamate, malate and ascorbic acid" [26, 38]. According to Bahadır and Sezgintürk [53], compared to earlier and present/modern considered biosensors in academic/research laboratories, the modern biosensors which are mostly commercialized are far further fewer indicatory of the truncated achievement rates in agri-food-connected biosensor development.

The limitations encumbering biosensor development in agriculture/food sector are substantial impediments, such problems are; "mass production, sensor lifetime, component integration and handling practicability" [38]. The motives behind these restrictions are that the utmost machineries applied in present and forthcoming agriculture/food biosensing technology are in their infancy/early stages and they include; "nanotechnology, agriculture/food material science, biomimetic chemistry and microengineering". These basic factors could assist in the determination of forthcoming biosensors industries is its safety to human well-being, which implies that it is those with limited or no human well-being effect will have their commercialization in the forthcoming years [38]. The commercialization of intelligent agriculture/food industry specifies urgent needs in new and effective procedures to guarantee food quality and safety, to economize production procedure and to diminish loss in agriculture [54].

Biosensors have been employed for the monitoring and management of remediation procedures via the determination of the parameters that influence the growth of microbes, such as nutrient accessibility, pH, metal ions, liquified oxygen and temperature [55]. Biosensors that are required for the detection of environmental contaminants on field or large scale are not difficult to handle and need little

Biosensors	Industries
BIACORE	Biacore AB located in Sweden
Model-Amp Biosens, BioITO biosens, SMAlgal	Biosensor srl located in Formello, Italy
MB-DBO, Polytox-Res, Biocounter	Biosensores SL located in Moncofar, Spain
Portable Toxicity Screen (PTS)	52 Biotechnology Ltd. located in Uxbridge, UK
Cellsense	Euroclon Ltd. located in Yorkshire, UK
DropSens- Screen printed Electrodes	DropSense located in Asturia, Spain
Model-B.LV5, Model-B.IV4	Innovative Sensor Technology located in Nevada, USA
NECi's Nitrate Biosensor	Nitrate Elimination Co. Inc. located in Michigan, USA
Optiqua EventLab™, Optiqua MiniLab™	Optisense located in Netherlands
REMEDIOS	Remedios located in Aberdeen, Scotland
SciTOX-ALPHA, SciTOX-UniTOX	SciTOX Ltd. located in Oxford, USA

Table 2.
Commercial biosensors industries for environmental monitoring and management.

volumes of sample rather than conventional analytical procedures which required active sample pre-treatment phases [25, 56, 57]. However, the quest of effective biosensors is continuously increasing, not just in the field of agricultural and environmental sciences, but also in other fields such as medical sciences and engineering. Presently, as a result of the wide range applications of biosensors, potential markets are still been advanced and some the few available commercial biosensors industries for environmental monitoring and management are listed in **Table 2**.

4. Conclusion

Climate smart agriculture purposes exceptional prospects for handling the issues of food security as well as easing the adaptation and mitigation succors for environmental and agricultural sustainability. Climate smart agriculture has been of great assistance in this regard to most developed nations. Implementing climate smart agriculture as a capable and swift climate change response is extremely vital for building capacity and achieving food security as well as sustainable agriculture and environment globally In developing nations especially those of Sub-Sahara Africa, viewing the susceptibility to the altering climatic/weather conditions, their substantial dependence on agriculture for livelihoods and the critical role agriculture play in their economic sector; they would predominantly benefit from climate smart agriculture. Considering these regions' susceptibility to the changing climatic condition, their heavy reliance on agriculture for livelihoods, and the critical position agricultural sector holds concerning food security in these nations climate smart agriculture would undoubtedly be of great assistance. Nevertheless, there is a necessity for variance methods in encouraging the acceptance and advancement of climate smart agriculture. The small-scale agricultural segment in most developing nations is categorized by a diverse inhabitant. Consequently, a solitary even method would not be suitable in advancing climate smart agriculture practices among these set of farmers. The consequence of this is that approaches to support climate smart agriculture implementation should factor in specific collective as a replacement for mainstreaming approaches globally. Consequently, all stakeholders should contemplate of employing modalities that can accommodate the diverse features of climate smart agriculture and circumvent the potential challenges that could otherwise ascend. Additionally, since climate smart agriculture development in developing countries depends on the willingness of those involved in agricultural activities, hence, there is a need for all stakeholders to understand the multi-dimensional climate change issues and the subsequent self-mobilization for evolving and executing strategies to respond to the issues at appropriate scales.

Conclusively, in spite of the numerous benefits of biosensors and biosensing machineries such as nanoparticles/nanomaterials, polymers and microbes built-biosensors in solving some of the challenges in agricultural activities vis-à-vis environmental sustainability; there is still the need to significantly assimilate multi-faceted methods in developing biosensors that can potentially be used for diverse applications in climate smart organic/biological agriculture for environmental sustainability. Therefore, it is suggested that appropriate combination of biosensing as well as bio-fabrication with non-natural/synthetic biology methods by applying either/both electrochemical, optical, bio-electronic moralities would be crucial for efficacious development of comprehensive and influential biosensors for contemporary future contribution to knowledge in the field of biosensor machinery in climate smart organic/biological agriculture for environmental sustainability.

Acknowledgements

The author is sincerely grateful to authors and agencies whose research studies and publications were used for this chapter.

Conflict of interest

There is no conflict to declare.

Notes

This research has not received any specific grant from funding agencies in the public, commercial, or not-for-profit sectors.

Author details

Kingsley Eghonghon Ukhurebor
Climatic/Environmental/Telecommunication Unit, Department of Physics, Edo University Iyamho, Edo State, Nigeria

*Address all correspondence to: ukeghonghon@gmail.com; ukhurebor.kingsley@edouniversity.edu.ng

References

[1] Garnett T. Food sustainability: Problems, perspectives and solutions. The Proceedings of the Nutrition Society. 2013;**72**:29-39

[2] Karimi V, Karami E, Keshavarz M. Climate change and agriculture: Impacts and adaptive responses in Iran. Journal of Integrative Agriculture. 2018;**17**(1):1-15

[3] Food and Agriculture Organisation of the United Nations (FAO). Climate Smart Agriculture Sourcebook. Rome, Italy; 2015

[4] Field CB, Barros VR, Dokken D, Mach K, Mastrandrea M, Bilir T, et al. IPCC: Climate Change 2014: Impacts, adaptation, and vulnerability. Part A: Global and sectoral aspects. In: Contribution of Working Group II to the Fifth Assessment Report of the Intergovernmental Panel on Climate Change. Cambridge, UK; New York, NY, USA: Cambridge University Press; 2014. pp. 1-32

[5] Akrofi-Atitianti F, Speranza CI, Louis Bockel L, Asare R. Assessing climate smart agriculture and its determinants of practice in Ghana: A case of the cocoa production system. Landscape. 2018;**7**:30

[6] Ukhurebor KE, Abiodun IC. Variation in annual rainfall data of forty years (1978-2017) for South-South, Nigeria. Journal of Applied Sciences and Environmental Management. 2018;**22**(4):511-518

[7] Herndon JM, Whiteside M. Further evidence that particulate pollution is the principal cause of global warming: Humanitarian considerations. Journal of Geography, Environment and Earth Science International. 2019;**21**(1):1-11

[8] Nwankwo W, Ukhurebor KE. An X-ray of connectivity between climate change and particulate pollutions. Journal of Advanced Research in Dynamical and Control Systems. 2019;**11**(8):3002-3011

[9] Nwankwo W, Ukhurebor KE, Aigbe UO. Climate change and innovation technology: A review. Kansai University. 2020;**63**(3):383-391

[10] Nwankwo W, Olayinka SA, Ukhurebor KE. Green computing policies and regulations: A necessity? International Journal of Scientific and Technology Research. 2020;**9**(1):4378-4383

[11] Ukhurebor KE, Azi SO. Review of methodology to obtain parameters for radio wave propagation at low altitudes from meteorological data: New results for Auchi area in Edo State, Nigeria. Journal of King Saud University – Science. 2019;**31**(4):1445-1451

[12] Ukhurebor KE, Olayinka SA, Nwankwo W, Alhasan C. Evaluation of the effects of some weather variables on UHF and VHF receivers within Benin City, South-South region of Nigeria. Journal of Physics: Conference Series. 2019;**1299**:012052

[13] Ukhurebor KE, Umukoro OJ. Influence of meteorological variables on UHF radio signal: Recent findings for EBS, Benin City, South-South, Nigeria. IOP Conference Series: Earth and Environmental Science. 2018;**173**:012017

[14] Ukhurebor KE, Nwankwo W. Estimation of the refractivity gradient from measured essential climate variables in Iyamho-Auchi, Edo State, South-South region of Nigeria. Indonesian Journal of Electrical Engineering and Computer Science. 2020;**19**(1):276-284

[15] Abegunde VO, Sibanda M, Obi A. The dynamics of climate change

adaptation in Sub-Saharan Africa: A review of climate-smart agriculture among small-scale farmers. Climate. 2019;7:132

[16] Fanen T, Olalekan A. Assessing the role of climate-smart agriculture in combating climate change, desertification and improving rural livelihood in Northern Nigeria. African Journal of Agricultural Research. 2014;**9**:1180-1191

[17] Antwi-Agyei P, Dougill AJ, Stringer LC. Barriers to climate change adaptation in Sub-Sahara Africa: Evidence from Northeast Ghana and systematic literature review. Climate and Development. 2014;7:297-309

[18] United Nations [Internet]. 2019. Available from: https://www.un.org/development/desa/en/news/population/world-population-prospects-2019.html [Accessed: 07 June 2020]

[19] Borgmann S, Schulte A, Neugebauer S, Schuhmann W. Amperometric biosensors. In: Alkire RC RC, Kolb DM, Lipkowski J, editors. Advances in Electrochemical Science and Engineering: Bioelectrochemistry. Vol. 13. Germany: Wiley-VCH Verlag GmbH & Co. KGaA; 2011

[20] Arora N. Recent advances in biosensors technology: A review. Octa Journal of Biosciences. 2013;**1**(2):147-150

[21] Vigneshvar S, Sudhakumari CC, Senthilkumaran B, Prakash H. Recent advances in biosensor technology for potential applications – An overview. Frontiers in Bioengineering and Biotechnology. 2016;**4**(11):1-9

[22] Salek-Maghsoud A, Vakhshiteh F, Torabia R, Hassani S, Ganjali MR, Norouzi P, et al. Recent advances in biosensor technology in assessment of early diabetes biomarkers. Biosensors & Bioelectronics. 2018;**99**:122-135

[23] El-Said WA, Abdelshakour M, Choi JH, Choi JW. Application of conducting polymer nanostructures to electrochemical biosensors. Molecules. 2020;**25**(307):1-11

[24] Turner APF, Karube I, Wilson GS. Biosensors Fundamentals and Applications. Oxford: Oxford University Press; 1987

[25] Rodriguez-Mozaz S, Alda MJ, Marco MP. Biosensors for environmental monitoring: A global perspective. Talanta. 2005;**65**(2):291-297

[26] Antonacci A, Arduini F, Moscone D, Palleschi G, Scognamiglio V. Commercially available (bio)sensors in the agrifood sector. In: Comprehensive analytical Chemistry. Vol. 74. Amsterdam, the Netherlands: Elsevier; 2016. pp. 315-340

[27] Reyes De Corcuera JI, Cavalieri JR. Prototype instruments for laboratory and on-line measurement of lipoxygenase activity. Food Science and Technology International. 2003;**9**(1):5-9

[28] Clark LC Jr, Lyons C. Electrode systems for continuous monitoring in cardiovascular surgery. Annals of the New York Academy of Sciences. 1962;**102**:29-45

[29] Fracchiolla NS, Artuso S, Cortelezzi A. Biosensors in clinical practice: Focus on oncohematology. Sensors. 2013;**13**:6423-6447

[30] Turner AP. Biosensors: Sense and sensibility. Chemical Society Reviews. 2013;**42**:3184-3196

[31] Citartan M, Gopinath SC, Tominaga J, Tang TH. Label-free methods of reporting biomolecular interactions by optical biosensors. The Analyst. 2013;**138**:3576-3592

[32] Sang S, Wang Y, Feng Q, Wei Y, Ji J, Zhang W. Progress of new label-free

techniques for biosensors: A review. Critical Reviews in Biotechnology. 2015;**15**:1-17

[33] Mehrotra P. Biosensors and their applications - A review. Journal of Oral Biology and Craniofacial Research. 2016;**6**(2):153-159

[34] Verma N, Bhardwaj A. Biosensor technology for pesticides – A review. Applied Biochemistry and Biotechnology. 2015;**175**:3093-3119

[35] Long F, Zhu A, Shi H. Recent advances in optical biosensors for environmental monitoring and early warning. Sensors. 2013;**13**:13928-13948

[36] Marrazza G. Piezoelectric biosensors for organophosphate and carbamate pesticides: A review. Biosensors. 2014;**4**: 301-317

[37] Adesina O, Anzai IA, Avalos JL, Barstow B. Embracing biological solutions to the sustainable energy challenge. Chem. 2017;**2**:20-51

[38] Neethirajan S, Ragavan V, Weng X, Chand R. Biosensors for sustainable food engineering: Challenges and perspectives. Biosensors. 2018;**8**:23

[39] Pundir CS, Chauhan N. Acetylcholinesterase inhibition-based biosensors for pesticide determination: A review. Analytical Biochemistry. 2012;**429**:19-31

[40] Senthilkumaran B. Pesticide- and sex steroid analogue-induced endocrine disruption differentially targets hypothalamo-hypophyseal-gonadal system during gametogenesis in teleosts – A review. General and Comparative Endocrinology. 2015;**219**:136-142

[41] Prasad R, Bhattacharyya A, Nguyen QD. Nanotechnology in sustainable agriculture: Recent developments, challenges, and perspectives. Frontiers in Microbiology. 2017;**8**:1014

[42] Dasgupta N, Ranjan S, Mundekka D, Ramalingam C, Shanker R, Kumar A. Nanotechnology in agro-food: From field to plate. Food Research International. 2015;**69**:381-400

[43] Duhan JS, Kumar R, Kumar N, Kaur P, Nehra K, Duhan S. Nanotechnology: The new perspective in precision agriculture. Biotechnology Reports. 2017;**15**:11-23

[44] Fraceto LF, Grillo R, Medeiros DAG, Scognamiglio V, Rea G, Bartolucci C. Nanotechnology in agriculture: Which innovation potential does it have? Frontiers in Environmental Science. 2016;**4**:20

[45] Liébana S, Lermo A, Campoy S, Cortés MP, Alegret S, Pividori MI. Rapid detection of Salmonella in milk by electrochemical magneto-immunosensing. Biosensors & Bioelectronics. 2009;**25**(2):510-513

[46] Pividori MI, Alegret S. Electrochemical genosensing based on rigid carbon composites. A review. Analytical Letters. 2005;**38**(15):2541-2565

[47] Brasil de Oliveira Marques PR, Lermo A, Campoy S, Yamanaka H, Barbe J, Alegret S, et al. Double-tagging polymerase chain reaction with a thiolated primer and electrochemical genosensing based on gold nanocomposite sensor for food safety. Analytical Chemistry. 2009;**81**(4):1332-1339

[48] Majumdar T, Agarwal S, Chakraborty R, Raychaudhuri U. Study on electrochemical immunoassay methodology for protein a: A modified approach towards detection and quantification of staphylococcus Aureus in food samples. Research Journal of Chemical Sciences. 2012;**2**:62-68

[49] Karsunke XY, Niessner R, Seidel M. Development of a multichannel

flow-through chemiluminescence microarray chip for parallel calibration and detection of pathogenic bacteria. Analytical and Bioanalytical Chemistry. 2009;**395**(6):1623

[50] Radke SM, Alocilja EC. A high-density microelectrode array biosensor for detection of E. coli O157: H7. Biosensors & Bioelectronics. 2005;**20**(8):1662-1667

[51] Kim N, Park IS. Application of a flow-type antibody sensor to the detection of Escherichia coli in various foods. Biosensors & Bioelectronics. 2003;**18**(9):1101-1107

[52] Papadakis G, Skandalis N, Dimopoulou A, Glynos P, Gizeli E. Bacteria murmur: Application of an acoustic biosensor for plant pathogen detection. PLoS One. 2015;**10**(7):e0132773

[53] Bahadır EB, Sezgintürk MK. Applications of commercial biosensors in clinical, food, environmental, and biothreat/biowarfare analyses. Analytical Biochemistry. 2015;**478**:107-120

[54] Vanderroost M, Ragaert P, Devlieghere F, De Meulenaer B. Intelligent food packaging: The next generation. Trends in Food Science and Technology. 2014;**39**:47-62

[55] Purohit HJ. Biosensors as molecular tools for use in bioremediation. Journal of Cleaner Production. 2003;**11**:293-301

[56] Iqbal SS, Mayo MW, Bruno JG, Bronk BV, Batt CA, Chambers JP. A review of molecular recognition technologies for detection of biological threat agents. Biosensors & Bioelectronics. 2000;**15**:549

[57] Mazzei F, Antiochia R, Botre F, Favero G, Tortolini C. Affinity- based biosensors in sport medicine and doping control analysis. Bioanalysis. 2014;**6**:225-245

Chapter 9

Efficacy of Risk Reducing Diversification Portfolio Strategies among Agro-Pastoralists in Semi-Arid Area: A Modern Portfolio Theory Approach

Ponsian T. Sewando

Abstract

Agro-pastoralists in the tropical semi-arid dryland areas of sub-Saharan Africa are significantly affected by climate change and variability. The agro-pastoral families are coping with production-related climatic risks through livelihood diversification to ensure food security. Data were collected from a sample of 411 agro-pastoralists across five districts in the semi-arid northern and central regions of Tanzania through survey conducted between November 2017 and July 2018. Secondary data regarding crop yields and livestock populations for eight years from 2009 to 2017 were collected from the National Bureau of Statistics and the respective District offices. Results show that about three-quarters of the agro-pastoralists managed diversified crop and livestock portfolios with two or more crops and animal species. However, simulated crop yields reveal positive correlations. Construction of integrated portfolios that generate good returns at a modest risk can be achieved through strategic choices between high-return high-risk and low-return low-risk crop and livestock activities. Thus, the paper recommends for costly long-term breeding and genotype improvement programs, strategically changing the make-up of the current crop and livestock portfolios which appear to be an affordable and tailored solution for building risk resilience among agro-pastoral communities in the drylands.

Keywords: modern portfolio theory, climate change and variability, farm returns, risks, variability

1. Introduction

Farming systems in sub-Saharan Africa (SSA) are especially vulnerable to natural hazards, as they are typically dependent on natural resources, such as limited and erratic rainfall with high inter- and intra-annual variability, pests and diseases, nutrient-poor soils and other natural calamities [1]. Weather and climate, with its inherent variability, means that farmers are facing risk, entailing either reduced or total production failure [2].

Effects of climate change on agriculture will be most adverse in regions that already suffer from high temperatures and low precipitation [3]. Such regions include the semi-arid drylands of SSA, where over three-quarters of the cropland is depends on the weather as it is rain-fed, hence further amplifying the sensitivity of agriculture to precipitation [3]. The expected increases in temperature for SSA is estimated to range from 2.0 to 4.5°C by 2100, while the annual rainfall for individual countries is expected to change by −39 to +64 mm by 2030 [4].

In order to cope with, and adapt to, climate change and variability, agro-pastoralists in the tropical areas of SSA are pursuing diversified livelihood strategies in crop and animal agriculture for enhancing food security. Crops are grown and livestock kept in diverse mixtures of varying sensitivity to production risks in order to ensure farm income and reduce the risk of failure [5]. Ref. [6] shows clearly that in semi-arid environments, combining crop production and livestock diversifies livelihoods. With limited public investments in planned adaptations, agro-pastoralists remain vulnerable to climatic risks while relying mostly on their autonomous coping strategies and adaptations.

The strategic diversification choices not only spreading risk [7], but also provide an important hedge against risk [8]. It is not prudent to invest all resources in highly correlated activities that may all perform poorly at the same time. Therefore, investing in two or more activities, whose returns are not full correlated, reduces the overall volatility below that of each one being taken separately [9]. As managers, agro-pastoralists guarantee food security by reducing the volatility of their farming by seeking a mix of farming activities that have either a small or negative correlation of related returns. This is critical for agro-pastoralists pursuing their livelihoods amid climatic risks are being further aggravated due to climate change.

Moreover, agricultural diversification enhances food security and farm income thus mitigating climate-related production risks [10]. Each crop-livestock combination has specific returns and risks. However, few studies evaluate the returns and risks associated with various crops and livestock portfolios among agro-pastoral farmers in the dryland areas of SSA.

This paper contributes to bridging this research gap by evaluating the levels of returns and risks associated with crop-livestock portfolios among agro-pastoralists in the semi-arid areas of Northern and Central Tanzania. The Modern Portfolio Theory (MPT) was used to evaluate risk in corporate and financial portfolio management, to evaluate the returns and risk of different crop-livestock portfolios which enhance food security. The results of this study inform the strategic diversification choices for enhanced resilience of agro-pastoralists in the face a changing climate.

2. Methodology

2.1 Theoretical framework

Smallholder agro-pastoral farmers are struggling to adapt to climate change and variability in order to maximize their utility (welfare) by safeguarding their livelihoods. The classical economic analysis of decision-making in the presence of risky and uncertain outcomes is based on Expected Utility (EU). EU theory underlies choices under risk [11]. EU theory provides a framework for modeling the choices of a rational individual whose goal is to maximize expected utility [12]. The underlying assumption of EU theory is that individuals have stable and coherent preferences; they know what they want and they examine the available alternatives before selecting the one that they judge to be best [13].

According to [12], rational choice theory is based on a model comprising two components: 1) a group of alternatives that are possible to realize, under different conditions; and 2) an individual's preferences that reflect their goals. However, the EU theory framework is criticized because it assumes that decision-makers are familiar with the probability distributions of each alternative outcome [12]. This is a serious theoretical flaw underlying EU in the face of climate change, which is inherently endowed with uncertainty [14, 15], not risk. Other complex alternative frameworks are suggested in the literature, including Subjective Utility Theory, Maxmin, α-maxmin, Minmax Regret, Maxmin EU, Smooth Ambiguity and Variational Preferences (Heal and Millner 2013). Kahneman and Tversky (1979) present a critique of expected utility theory as a descriptive model of decision making under risk and develop an alternative model, which they call *prospect theory*.

Despite its shortcomings, EU is still a useful theoretical framework for understanding the choices of agro-pastoral household livelihood activities. This is because agro-pastoralists in the semi-arid tropical drylands have lived with the vagaries of weather extremes for generations. In this regard, they have learned the patterns of climatic risks that impact their livelihood activities, meaning that risks due to climate change will not be a completely new experience.

Extremely risky situations, such as natural disasters, like climate extremes, have a significant effect on the probability distributions, resulting in the tails of the distribution being emphasized. On the other hand, investment alternatives, such as adaptations, are determined by the shape and symmetry of the expected outcome distributions [12]. Markowitz's method of optimization [16] in the context of Modern Portfolio Theory (MPT), can be adapted to include anomalies, heavy-tailed and asymmetric distributions, as well as more sophisticated measures of extreme risks. These technical attributes support its use in risk analysis in riskier undertaking such as crop and animal agriculture in the dryland farming systems.

In this paper, we extend EU theory into the MPT framework in order to analyze alternative portfolios of crop and livestock types by agro-pastoralists in terms of their returns and associated risks. MPT uses information about the joint probability distribution of outcomes of all possible assets in a portfolio (including means, variances, and covariances) in order to select a portfolio that efficiently manages risk [17]. Although MPT has been applied to financial and corporate portfolio management since the 1950s, its potential for analyzing climate change adaptation and resilience by evaluating returns of livelihood activities and associated risk remains unfulfilled.

The MPT framework is increasingly adapted for risk management in agriculture, in particular focusing on diversification in agriculture and forestry for reasonably higher and stable output in the face of uncertain climate [1, 3, 18]. There are three assumptions of MPT for diversification: (i) there is more than one possible investment at any given time; (ii) these investments are subject to risk; and (iii) that the same economic and environmental conditions do not affect all investments equally [19].

The livelihood resilience of agro-pastoralists in the face of climate change and variability would be enhanced if crop-livestock portfolios generate upgraded returns with minimal variances. Optimal diversification by combining activities with low positive covariance and income-skewing is a primary risk reducing strategy. This is achieved by reducing the risk of the overall return by selecting a mixture of activities whose net returns have a low or negative correlation [20]. This means that farmers spread risk by diversifying the allocation of productive assets across various income-generating activities, often preferring farm plans that provide a satisfactory level of security even if this means sacrificing income on average [21].

Agro-pastoralists might wish to avoid income fluctuation and to maximize income at the same time. Since farms can be thought of as assets within an overall portfolio, agricultural producers might behave in a manner like investors who pay attention to the concept of diversification. However, agro-pastoralists might not have considered diversification in the same way that the typical financial investor might: they often look at diversification as changing their crop mix, rotational system, and livestock breeds, or even as cultivating spatially separated farms [22] and splitting livestock across distant grazing locations [23]. Despite the relevance of considering these forms of diversification, our analytical scope is limited to diversification of alternative combinations of crop and livestock enterprises. However, it highlights possible future research for increasing the scope of portfolio diversification in SSA.

2.2 The study area

The study was conducted in the semi-arid drylands of northern and central Tanzania, covering five districts namely Mwanga, Arusha, Babati, Kongwa and Ikungi from Kilimanjaro, Arusha, Manyara, Dodoma and Singida regions, respectively. The long-term temperatures and rainfall (1990–2016) were analyzed by considering averages of five base years from 1990 to 1995 and the five years of 2009–2016. The aim was to investigate if there was a notable shift in temperatures and rainfall in over the long-term. Unlike comparing a single base (1990) and terminal year (2016), the clusters of five years in the base and terminal periods facilitate capturing intermittent annual volatility.

The average annual monthly minimum temperature has increased by around 1°C in northern Tanzania (**Table 1**). This means that over time, months that used to be cooler are getting warmer. The maximum temperature has risen by 0.3°C in Arusha and 0.9°C in Mwanga. The standard deviations are larger, thus indicating increased volatility of temperatures. Rainfall has only increased marginally in Mwanga and Babati (+16.3 and + 1.3 mm, respectively) and decreased by 72.2 mm in Arusha. The magnitudes of standard deviations indicate higher inter-rainfall annual variability.

In the central semi-arid areas, in the cooler months, Kongwa has become warmer by about 1°C, while warmer months have registered an average increase of about 0.4°C over time (**Table 1**). Ikungi has registered marginally decreasing (−0.3°C) minimum and increasing (+0.1°C) maximum temperatures. In terms of

Study locations	**Temperature (°C)**				**Rainfall (mm)**	
District	Base years [earliest 5 years[ψ]]		Now [recent 5 years[†]]		Base years [earliest 5 years[ψ]]	Now [recent 5 years[†]]
	Min	**Max**	**Min**	**Max**		
Mwanga	17.0 (0.32)	29.9 (0.82)	18.1 (0.36)	30.7 (0.34)	431.9 (146.7)	448.2 (80.9)
Arusha	14.4 (0.26)	26.0 (0.61)	14.9 (0.29)	26.3 (0.47)	775.1 (195.9)	702.8 (132.7)
Babati	na	na	na	Na	584.4 (183.3)	585.7 (227.9)
Kongwa	16.8 (0.40)	29.0 (1.28)	17.6 (0.14)	29.4 (0.19)	517.2 (30.0)	566.6 (168.5)
Ikungi	16.5 (0.58)	27.5 (0.38)	16.2 (0.17)	27.6 (0.34)	608.5 (188.8)	668.8 (193.9)

Numbers in open and closed brackets are averages and standard deviations, respectively; na = data not available.
[ψ]Base earliest 5 years between 1990 and 1995.
[†]The 5 years between 2009 and 2016.

Table 1.
Average monthly annual temperature and rainfall trend in the study areas.

rainfall, locations in central Tanzania have experienced an increase of monthly annual average rainfall between 50 and 60 mm.

Generally, over time, the study locations are getting warmer and experiencing an increase in rainfall. However, larger standard deviations, particularly with rainfall, indicate higher variability. Increasing temperatures will counteract marginal increases in rainfall through increased evapo-transpiration, thus inducing stress on plant and animal production. While our results highlight the general trend in temperature and rainfall, a more rigorous analysis of rainfall distribution in terms of the number of rainy days would provide more information on the rainfall distribution, which is what matters most for agricultural production.

2.3 Research design and sampling

This study used a cross-sectional survey design that allowed data to be collected at a single point in time, while still having a broad scope. The study used crop livestock panel data over the period of 8 years (2009–2017), available at the district level, are also used in the analysis. A simplified formula for proportions by [8] is adopted in order to obtain the desired sample size of agro-pastoralists in semi-arid northern and central Tanzania, assuming 95% confidence level and 0.05 as sampling error; the formula is as follows in Eq. 1:

$$n = \frac{N}{1 + N\left(e^2\right)} \tag{1}$$

where n is the sample size; N is the population size; and e is the level of precision i.e. sampling error.

Using the above formula, a total of 411 agro-pastoral households were sampled.

A multi-stage sampling approach was employed in order to select the study locations and, ultimately, the individual agro-pastoral households. The study locations from region, district, ward, and down to the village were selected based on the criteria of having the largest number of agro-pastoralists. The study covered 5 regions, each represented by one district, as well as an overall total of 12 wards and 23 villages. The last stage was random selection of agro-pastoral farmers from lists of heads of agro-pastoral households as provided by agricultural field officers in the selected sample villages.

2.4 Data types and data collection

Both primary and secondary data are used in this study. Primary data were collected through structured household questionnaire interviews. The questionnaire covered among other information, the production costs, yields and sales. Secondary data were collected from respective offices at the meteorological weather stations, the National Bureau of Statistics (NBS) online database, and district offices. The secondary data included historical crop yields and livestock numbers, with the latter expressed in livestock units (LU). The conversion of different livestock species into a standard LU was based on the coefficients suggested by [24]: one animal representing 0.7, 0.1, 0.1 and 0.01 LU for cattle, goat, sheep, and chickens, respectively.

2.5 Data analysis

2.5.1 Adapted definitions of MPT terminologies

Beforehand, it was important to present the conventional MPT terminologies and how they were applied in this paper (**Table 2**).

Terminology	Usage in financial investments and in this paper
Asset/securities	Items within a portfolio which refers to crops and livestock species kept by the agro-pastoralist.
Correlation	A measure of the degree to which the change in the return of two assets is similar. It represents correlations of crop and livestock activities.
Diversification	Investing in different assets that together make up a portfolio. It refers to crop and livestock activities.
Efficient portfolio	A portfolio offering maximal expected return at a chosen level of risk or minimal risk at a given return. In this study, it is the crop and livestock portfolio giving relatively higher returns at given levels of risk.
Portfolio	Set of financial assets held by an investor. It refers to the crop and livestock combinations pursued by the agro-pastoralist.
Return	Financial returns from a financial investment asset. The concept refers to the value of crop outputs and livestock units managed by the agro-pastoralist.
Risk	The uncertain outcome of a financial investment expressed as standard deviations or variance. It refers to the standard deviations associated with returns from crop and livestock portfolios.

Table 2.
Terminologies of the MPT as applied in this paper.

2.5.2 Incorporating local management into district crop yields and livestock units

The cross-sectional survey used in our study cannot generate longer time series data because even recalling data from five-years before is problematic as smallholder farmers do not keep records. In order to predict future (expected) returns for crop and livestock portfolios, frequently the historical performance of respective returns are examined [25, 26]. The secondary data for eight years (2009–2017) are the longest time series of crop yields and livestock numbers available from the study districts. These secondary data were the basis for simulating the distributions of future expected yields and livestock units.

Given the nature of data collection and management systems of family farms, the aggregated district-level data can have biases and errors resulting from the aggregation process. Common crop yield estimates are often based on rough estimates of aggregate production and area harvested. Owing to significant variation in farming practices and growing conditions across farming systems and agro-ecological zones, higher-level yield estimates may differ starkly from local yields realized by any given smallholder farmer [27]. According to [28], yields and related economic returns vary overtime and space, with this variation important for understanding the vulnerability of farms to climatic risks. Therefore, we used the cross-sectional household survey crop yields and livestock units to normalize respective district-level data. The normalization tends to localize aggregated data thus improving the reliability and validity of yields and livestock units at the local scale. The normalization approach is adapted from the work of [28] as follows in Eq. 2.

$$\beta_{aij} = y_{aijk} / y_{hik} \tag{2}$$

where y_{aijk} = aggregate district-level yield/livestock specie unit of ith crop/livestock type in j^{th} year of the reference production period (2009–2017), in k^{th} district; y_{hik} = average observed household yield/livestock type unit of ith crop/livestock type in k^{th} district; β_{aij} = normalization factor of aggregate district level crop yield/livestock type unit of ith crop/livestock type in j^{th} district.

The factor is used to downscale secondary aggregated crop yields and livestock units to the local context. Thus, the normalized crop yields and livestock units were used to derive returns and associated risks as shown in Eq. 3.

$$y_{nij} = \beta_{aij} \times y_{aijk} \tag{3}$$

where y_{nij}= normalized aggregate district-level yield/livestock type unit of ith crop/livestock type in j^{th} year of reference production period.

2.5.3 Expected returns of crop or livestock

The size of the returns was measured by the expected mean value of the distribution. The expected mean return $E(R)$ is given by the weighted average of all possible returns, using the probability of achieving the actual return (R_i). However, this study employs historical data, meaning that the expected or mean return R_i of an individual agro-pastoral return is derived from normalized crop yields and livestock units during the 2009–2017 period. The expected returns of crops and livestock are estimated as shown in Eq. 4:

$$\overline{R_i} = \sum_{s=1}^{n} p_s R_s \tag{4}$$

where $\overline{R_i}$= expected return of crop enterprise per hectare over the 2009 to 2017 period of a particular crop and livestock units per household over the 2009 to 2017 period; R_s = historical actual returns, and n = number of years.

2.5.4 Risk of individual agro-pastoralist activities

The risk of an individual agro-pastoral's crop and livestock portfolios was measured using respective standard deviations. For most agro-pastoralists as "investors," risk is experienced when they engage in crop and livestock production activities that generate returns that are lower than what was expected. As a result, it is a negative deviation from the expected (average) return. In other words, each crop and livestock activity presents its own standard deviation [26]. A higher standard deviation translates into greater risk. The standard deviation of a return is the square root of the variance. In general, the variance can be calculated as shown in Eq. 5:

$$\sigma^2 = \left[\sum_{i=1}^{n} \left(R_i - \overline{R}\right)^2\right] \div (n) \tag{5}$$

where σ^2= variance, n = number of year, R_i = historical actual return, and $\overline{R}$= expected returns.

2.5.5 Expected returns of a portfolio

Regardless of how the individual return was calculated, the expected return of a portfolio is the weighted sum of the individual returns from the crops and livestock that form portfolio, as presented in Eq. 6:

$$\overline{R}_p = \sum_{s=1}^{n} w_s \overline{R}_s \tag{6}$$

Where: w_s = the proportion of the value of the portfolio constituted by the current market value of the ith crops or livestock entities, that is, it is the 'weight' attached to the crop or livestock, n = the number of crops or livestock in the portfolio, $\overline{R}_s$ = historical actual return, $\overline{R}_p$= expected return of the portfolio.

2.5.6 Risk associated with portfolio diversification strategy

Risk is the chance that an investment's return will be different from what is expected. This includes the possibility of losing some or all of the original investment. Modern Portfolio Theory [16] offers a solution to the problem of portfolio choice for a risk-averse investor like an agro-pastoralist. The optimal portfolios, from the rational investor's point of view, are defined as those having the lowest risk for a given return. These portfolios are said to be mean–variance efficient [29]. Portfolio risk is measured by calculating the standard deviation of the historical returns or average returns of a specific investment. The standard deviation of the portfolio's rate of return depends on the standard deviations of return for its entities, their correlation coefficients, and the proportions invested. It is calculated as shown in Eq. 7:

$$\sigma_p = \sqrt{\sum_{i=1}^{n}\sum_{i=1}^{n} X_i X_j \rho_{ij} \sigma_i \sigma_j} \tag{7}$$

where σ_p = standard deviation of the portfolio,

$X_i X_j$ = proportion invested in each asset,

ρ_{ij} = correlation coefficients between i and j,

$\sigma_i \sigma_j$ = standard deviation of each asset, and n = number of years.

2.6 Results and discussion

2.6.1 Crops and livestock diversification in Tanzania

Diversification is widely used strategy to spread risk as crops differ in their sensitivity to climate and weather extremes. About three-quarters (77%) of the responding agro-pastoral households diversified with two or more crops. A substantial number of agro-pastoral households only grow maize (**Figure 1(a)**). The eight widely diversified crop portfolios, with at least ten counts of involved agro-pastoral households, included maize-sunflower, maize-cowpea, maize-groundnut, maize-millet-sunflower, maize-pigeon pea, maize-millet, maize-sorghum, and maize-green gram. Maize as the main preferred staple features in all major crop portfolios. With exception of maize, which is relatively sensitive to seasonal droughts and intermittent dry spells, the other crops tolerate drought, a characteristic of the semi-arid farming environment. This means that agro-pastoralists in these dryland areas tend to make a strategic choice of crop mixes in order to minimize production risks. The high preference of maize as a food staple suggests that there must be efforts in crop breeding and improvement programs in order to develop drought efficient maize varieties.

Diversification involving two or more livestock types involved around 70% of the sampled agro-pastoralists. Chickens were kept by the majority of the agro-pastoralists (**Figure 1(b)**). The prominent livestock diversification portfolios included cattle-goat-chicken, cattle-goat, cattle-chicken, and goat-chicken. Despite the fact that cattle production is highly vulnerable to seasonal droughts drying up pasture and water sources, it remains a part of the livelihood activities of agro-pastoral communities. Cattle are part of the social identity of

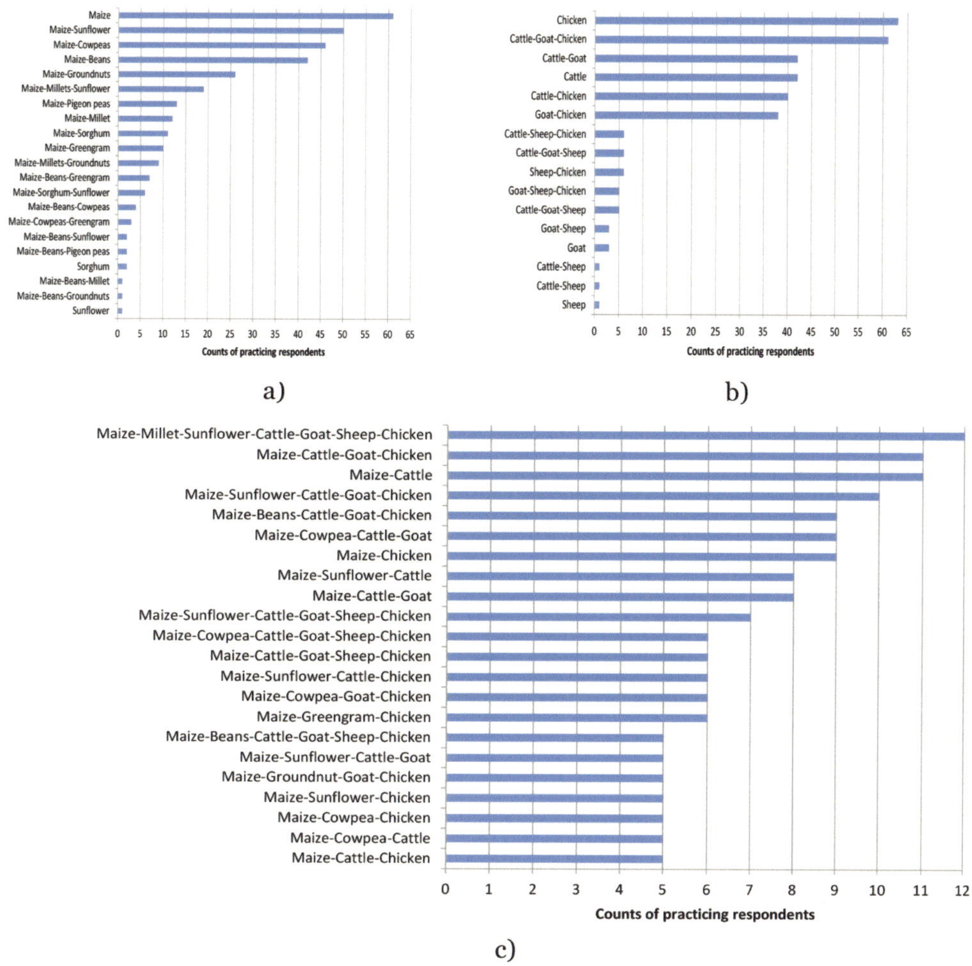

Figure 1.
Current diversification of crop and livestock portfolios. (a) Crop portfolio diversification. (b) Livestock portfolio diversification. (c) Crop-livestock portfolios diversification.

agro-pastoral societies in Tanzania. Goats, appearing in every major diversified livestock portfolio, are well adapted to the drier environment due to its ability to browse on trees and shrubs. Scavenging rural chickens can also thrive through critical drought periods.

Furthermore, diversification strategies integrating crops and livestock stand a better chance of effectively spreading production risk. Four major used diversified crop-livestock portfolios with at least ten counts of respondent agro-pastoralists include maize-millet-sunflower-cattle-goat-sheep-chicken, maize-cattle-goat-chicken, maize-cattle, and maize-sunflower-cattle-goat-chicken. Maize also features in the integrated crop-livestock portfolios. Goat and chickens, which are relatively less affected by droughts, were part of the integrated portfolios. Goats can browse on shrubs when the land lacks grazing grasses due to seasonal droughts. Cattle and sheep, which are reliant upon grasses, are highly vulnerable when grazing grounds dry up due to prolonged seasonal droughts. The prominence of cattle that are less-resilient to critical droughts and stressful heat urges for interventions to reduce risk in cattle production. Such interventions include water harvesting and storage, as well as climate-smart breeding and genotype improvement for adaptability to the drier environment.

2.6.2 Crop and livestock production with associated risks

2.6.2.1 Localized crop yields with associated risks

The district-level secondary data for seven growing seasons were localized through normalization with observed crop yields and livestock units in the survey localities. **Table 3** shows that, exception for millet, localized yields were higher than unadjusted aggregate district-level crop yields. For most crops, including sorghum, green gram, maize, groundnuts and beans, the localized yields ranged between a quarter and a half percent over the aggregate district-level yields.

The temporal volatility of locally adjusted crop yields, as reflected by the coefficients variation, was apparently higher than unadjusted district-level yields. Implicitly, local yields were highly variable with higher variances that were extended into adjusted aggregate yields through the normalization process. Although the aggregate data includes data collected outside the survey localities, it is still plausible to argue that the study locales represent the areas of respective districts. Therefore, in order to realistically reflect the local production risks faced by agro-pastoralists, downscaling higher level information is important. Furthermore, crop yields for most crops were below the national average of 1–2 tons for most for cereals and grain legumes. This reflects the higher production risks associated with crop production in semi-arid dryland areas.

2.6.2.2 Localized livestock units with associated risk

Results in **Table 4** indicate that cattle are the most significant livestock asset. Locally adjusted livestock units tended to be relatively smaller for cattle and sheep. There was some increase in the locally adjusted livestock units for goats and chickens. Given that the local livestock units were the denominator of the normalization factor, it means there were fewer goats and chicken but more cattle and sheep in the locales on average than in the rest of the respective districts. At the district scale, the numbers of goats and chickens were much variable over the period of 8 years. Locally adjusted livestock units tended to vary widely over the years, but with only limited differences across animal types. Goats and chickens are normally sold by agro-pastoralists for immediate cash needs, while decisions to sell cattle

Crops	Unadjusted district-level revenue (ton/ha)			Locally adjusted revenue (ton/ha)		
	Mean	Std. Dev.	CV	Mean	Std. Dev.	CV
Maize	265155.20	63638.00	0.24	375812.00	245828.80	0.65
Cowpeas	482332.80	352575.20	0.73	511172.00	507919.60	0.99
Beans	283974.00	89206.00	0.31	578570.00	466446.80	0.81
Sorghum	180968.80	36960.80	0.2	245302.40	136375.20	0.56
Green gram	201761.60	140492.40	0.7	335786.80	386753.60	1.15
Millets	245716.00	78527.60	0.32	235958.80	113006.80	0.48
Sunflower	179408.40	74166.00	0.41	229247.20	166135.60	0.72
Pigeon peas	70556.40	70255.60	1	125584.00	194655.20	1.55
Groundnuts	834024.40	564639.20	0.68	1395787.20	1583467.60	1.13

Table 3.
Unadjusted and locally adjusted revenue (Tshs).

Types	Unadjusted livestock			Locally adjusted livestock		
	Mean	**Std. Dev.**	**CV**	**Mean**	**Std. Dev.**	**CV**
Cattle	2449339.20	852204.00	0.35	750101.20	1108561.00	1.48
Goats	200464.40	141244.40	0.7	258387.20	383990.00	1.49
Sheep	103268.40	43691.20	0.41	18273.60	22033.60	1.21
Chicken	71740.80	56381.20	0.76	127069.20	150926.40	1.19

Table 4.
Unadjusted and locally adjusted livestock revenue (Tshs).

are carefully considered. Therefore, due to the volatility in the local production conditions and outputs, the aggregated information cannot be generalized for local realities.

2.6.2.3 Potential diversification of crop and livestock portfolios

An efficient portfolio is either a portfolio that offers the highest expected return at a given level of risk, or one with the lowest level of risk for a given expected return. The majority of agro-pastoralists manage combinations of two crops and two livestock species. The paired combinations of crops and livestock were the basis for developing integrated portfolios through a permutation process. **Table 5** shows one crop-livestock portfolio (Sorghum, sunflower, cattle and goat) with highest returns at a less level of risk. As presented earlier, maize is the predominant enterprise. Maize-beans-cattle-sheep had the highest returns followed by millet-beans-cattle-chickens.

Portfolio	Expected return (TShs)	Risks (TShs)	CV
Maize, beans, cattle and sheep	382937.20	206574.40	0.54
Maize, sorghum, cattle and chicken	442307.60	221783.60	0.5
Millet, beans, cattle and chicken	240376.80	140229.20	0.58
Maize, groundnut, cattle and chicken	330034.00	175310.00	0.53
Maize, cowpea, cattle and goat	283015.20	157092.80	0.56
Maize, sunflower, cattle and goat	283541.60	156077.60	0.55
Sorghum, sunflower, cattle and goat	1129880.00	486694.40	0.43
Maize, pigeon pea cattle and sheep	140962.40	130660.00	0.93
Maize, green gram, goat and sheep	108776.80	106370.40	0.98
Maize, millet, cattle and goat	386227.20	204694.40	0.53

Table 5.
Expected returns and risks of potential crop-livestock portfolios.

Apparently, the high-return high-risk portfolio categories tended to include cattle and beans, which are high value commodities. However, these two commodities are sensitive to climatic shocks, especially droughts. Beans are the major source of food protein that are widely consumed and traded in both rural and urban areas of Tanzania. Beans from northern Tanzania are also exported to Kenya. Maize-green gram-goat-sheep and maize-millet-cattle-goat were the least-risk, lowest-return portfolios. Early maturing maize varieties, millet, and green gram are drought tolerant and, hence, risk efficient. Likewise, goats are relatively less vulnerable than cattle. However, the downside risk associated with cattle is downplayed when this activity is part of the risk-efficient crop-livestock portfolios.

2.7 Conclusions and policy implications

Agro-pastoral households in the semi-arid areas of central and northern Tanzania have lived with the vagaries of weather, coping and adapting through an array of diversified crop and livestock portfolios. In the face of a changing climate, along with other production and market risks, current and potential crop and livestock portfolios must be empirically illuminated in order to test their underlying efficacy in ensuring acceptable returns and associated risks. Shedding light on the levels of returns and associated risks is critical in informing the future design of crop-livestock portfolios that enhance the livelihood resilience of agro-pastoral households in semi-arid areas of Tanzania.

The majority of agro-pastoralists are diversifying within and integratively across crop and animal farming activities. Some crops and livestock species, such as maize, cattle and sheep, are sensitive to climatic related production risks, but are widely raised. Apart from long-term breeding and genotype improvement programs to develop varieties and breeds that are drought resistant, the strategic reorientation of crop and livestock portfolios appears to be an affordable and tailored solution. For instance, goats and chickens that can thrive under critical drought conditions should be promoted in order to create portfolios that generate acceptable returns while minimizing risk.

Crop and livestock productivity in the semi-arid areas of central and northern Tanzania are generally low compared the national averages. Factors that contribute to such lower productivity among others include dependence on weather-dependent rainfed agriculture and lack of rainwater harvesting infrastructure. However, semi-arid Tanzania is home to a majority of the poor, with destitute cropping and herding families, thus making them poverty hotspots where poverty reduction efforts should be strategically targeted. Our analyses suggest an array of crop and livestock portfolios that can be promoted in order to generate economic returns within certain bounds of risks.

This study investigates the levels of economic returns and associate risks for the current and potential crop and livestock portfolios. In order to highlight the centrality of strategic choices of economic portfolios among agro-pastoralists for enhanced livelihood resilience, it is important to incorporate off-farm activities into the portfolio diversification analysis. However, this is left as an empirical niche for further studies.

Author details

Ponsian T. Sewando
Tengeru Institute of Community Development (TICD), Arusha, Tanzania

*Address all correspondence to: ponsiansewando@gmail.com

References

[1] Werners, S. E., Incerti, F., Bindi, M., Moriondo, M. and Cots, F. (2007). Diversification of Agricultural Crops to Adapt to Climate Change in the Guadiana River Basin [Internet]. 2007. Available from: http://www.hkccf.org/download/iccc2007/30May/S4A/Saskia%20WERNERS/Diversification%20of%20Agricultural%20Crops%20to%20Adapt%20to%20Climate%20Change%20in%20the%20Guadiana%20River%20Basin.pdf [Accessed: 2018/11/25

[2] Aw-Hassan, A., Rida, F., Telleria, R. and Bruggeman, A. The impact of food and agricultural policies on groundwater use in Syria. Journal of Hydrology, 2014; 513: 204-215. DOI.org/10.1016/j.jhydrol.2014.03.043

[3] Bodin, P., Olin, S., Pugh, T. A. M. and Arneth, A. Optimizing cropland cover for stable food production in Sub-Saharan Africa using simulated yield and Modern Portfolio Theory. Earth System Dynamic Discuss 2014; 5: 1571-1606. DOI.org/10.5194/esdd-5-1571-2014

[4] Jarvis, A., Ramirez-Villegas, J., Herrera Campo, B.V. and Navarro-Racines, C. (2012). Is cassava the answer to African climate change adaptation? Tropical Plant Biology. 2012; 5(1): 9-29. DOI: 10.1007/s12042-012-9096-7

[5] Kahan, D. Managing Risk in Farming. Food and Agricultural Organization, Rome, Italy. 2008. Available from: http://www.fao.org/uploads/media/3-ManagingRiskInternLores.pdf [Accessed: 2018/02/14

[6] Kangalawe, R. Y. M., and Lyimo, J. G. (2013). Climate change, adaptive strategies and rural livelihoods in Semiarid Tanzania. Natural Resources. 2013; 4: 266-278. DOI: 10.4236/nr.2013.43034

[7] Fabozzi, F. J., Focardi, S. M. and Kolm, P. N. Quantitative Equity Investing: Techniques and Strategies. John Wiley and Sons Inc., New Jersey; 2010. 511 p.

[8] Yamane, T. Statistics: An Introductory Analysis, 2nd Edition, New York: Harper and Row. 1967

[9] Kemuma, M. C. Effect of investment decision on the performance of firms listed in the Nairobi securities exchange [Dissertation]. University of Nairobi, Kenya, 2014.

[10] Lynn, S. The Pastoral to Agro-Pastoral Transition in Tanzania: Human Adaptation in an Ecosystem Context [Internet]. Available from: https://www.researchgate.net/publication/265309217 [Accessed: 209/04/22]

[11] von Neumann, J., and Morgenstern, O. Theory of games and economic behaviour, 2nd Ed. Princeton, NJ: Princeton University Press. 1947. 625 p.

[12] Stanković, J. and Petrović, E. Expected utility theory under extreme risks. FACTA UNIVERSITATIS Series: Economics and Organization, 2016; 13(1): 31-44 [Internet]. Available from: http://casopisi.junis.ni.ac.rs/index.php/FUEconOrg/article/view/1714/1118.pdf [Accessed: 2019/06/01

[13] Hedesström, T. M. The Psychology of Diversification: Novice Investors' Ability to Spread Risks. Department of psychology, Göteborg University, Sweden; 2006.

[14] Kunreuther, H., Heal, G., Allen, M., Edenhofer, O., Field, C.B., and Yohe, G. Risk management and climate change, Nature Climate Change, 2014; 3: 447-450. DOI: org/10.1038/nclimate1740

[15] Kunreuther, H. and Weber, E. U. Aiding decision-making to reduce

the impacts of climate change. Working. Journal of Consumer Policy, 2013; 37(3): 397-411. DOI: 10.1007/s10603-013-9251-z

[16] Markowitz, H. Portfolio selection. The Journal of Finance. 1952; 7(1): 77-91. DOI: 10.2307/2975974

[17] Ando, A. W. and Mallory, M. L. Optimal portfolio design to reduce climate-related conservation uncertainty in the Pr*airie Pothole Region.* Proceedings of the National Academy of Sciences of the United State of America (PNAS), Stanford University, Stanford, 23 February, 2012. DOI.org/10.1073/pnas.1114653109.

[18] Crowe, K.A. and Parker, W.H. Using portfolio theory to guide reforestation and restoration under climate change scenarios. Climatic Change,2007; DOI: 10.1007/s10584-007-9373-x.

[19] Fraser, E. D. G., Mabee, W. and Figge, F. A framework for assessing the vulnerability of food systems to future shocks. Futures. 2005; 37: 465-479. DOI: 10.1016/j.futures.2004.10.011

[20] Di Falco, S. and Chavas, J. P. (2009). On crop biodiversity, risk exposure, and food security in the highlands of Ethiopia. American Journal of Agricultural Economics. 2009; 91(3): 599-611. DOI:. org/10.1111/j.1467-8276.2009.01265.x

[21] Witt, R. and Waibel, H. Constraints to diversification of poor fishery-dependent households in Cameroon. The African Journal of Agricultural and Resource Economics, 2011; l6(2): 155-175. DOI: 10.22004/ag.econ.156972

[22] Libbin, J. D., Kohler, J. D. and Hawkes, J. M. Does modern portfolio theory apply to agricultural land ownership? Concepts for farmers and farm managers. Journal of the American Society of Farm Managers and Rural Appraisers, 2004; 85-96 [internet]. Available from: http://citeseerx.ist.psu.edu/viewdoc/download?doi=10.1.1.525.7166&rep=rep1&type=pdf [Accessed: 2019/07/03]

[23] Tumbo, S., Mutabazi, K., Kimambo, A. and Rwehumbiza, F. Costing and planning of adaptation to climate change in animal agriculture in Tanzania. International Institute for Environment and Development (IIED), London, UK [Internet]. 2011. Available from: http://www.seachangecop.org/sites/default/files/documents/2011%2008%20IIED%20%20Planning%20and%20costing%20agricultural%20adaptation%20to%20CC_Tanzania.pdf [Accessed: 2019/01/25

[24] Harvest Choice (2011). Total livestock population. International Food Policy Research Institute, Washington DC. Available from: http://harvestchoice.org/node/4788.html. Accessed: 2016/3/28.

[25] Mangram, M. E. A simplified perspective of the Markowitz Portfolio Theory, Global Journal of Business Research. 2013; 7(1): 59-70 [Internet]. Available from: https://www.researchgate.net/publication/256034486_A_Simplified_Perspective_of_the_Markowitz_Portfolio_Theory.pdf [Accessed: 2019/04/10]

[26] McClure, B. (2010). Modern portfolio theory: Why it's still hip. Investopedia [Internet]. Available from: http://www. investopedia.com/articles/06/MPT.asp#axzz1g3JQY7nY] site visited on 2019/10/20

[27] Reynolds, T. W., Anderson C. L., Slakie, E. and Gugerty, M. K. How Common Crop Yield Measures Misrepresent Productivity among Smallholder Farmers. A paper presented at the International Conference of Agricultural Economists, 8-14 August 2015, Milan, Italy. 33 p.

[28] Antle, J. and Valdivia R. Calculating Statistics for Climate Impact

Assessments Using Crop/Livestock Model Simulations, RAPs and the TOA-MD Model. In: The agricultural Model Inter comparison Project AgMIP Protocol for AgMIP Regional Integrated Assessment, 2015; Version 6.

[29] Leung, A.H. Portfolio selection and risk management: An introduction, empirical demonstration and r-application for stock portfolios. [Dissertation] University of California, Los Angeles, 2009.

Chapter 10

Arabian Sea Tropical Cyclones: A Spatio-Temporal Analysis in Support of Natural Hazard Risk Appraisal in Oman

Suad Al-Manji, Gordon Mitchell and Amna Al Ruheili

Abstract

Tropical cyclones [TCs] are a common natural hazard that have significantly impacted Oman. Over the period 1881–2019, 41 TC systems made landfall in Oman, each associated with extreme winds, storm surges and significant flash floods, often resulting in loss of life and substantial damage to infrastructure. TCs affect Omani coastal areas from Muscat in the north to Salalah in the south. However, developing a better understanding of the high-risk regions is needed, and is of particular interest in disaster risk reduction institutions in Oman. This study aims to find and map TC tracks and their spatio-temporal distribution to landfall in Oman to identify the high-risk areas. The analysis uses Kernel Density Estimation [KDE] and Linear Direction Mean [LDM] methods to better identify the spatio-temporal distribution of TC tracks and their landfall in Oman. The study reveals clear seasonal and monthly patterns. This knowledge will help to improve disaster planning for the high-risk areas.

Keywords: Oman, Arabian Sea, tropical cyclone, storm track analysis, natural hazard risk assessment, hazard mapping

1. Introduction

Understanding and forecasting the consequences of climate change is critical to support the work of planners and decisions makers. Climate change has touched many aspects of our environment and life; for example, 43% of all natural disasters are related to flooding, which has impacted 56% of all people around the world [1]. Rising global temperature drives more frequent and extreme rainfall events [2], and through thermal expansion, sea level rise. It has been reported that the temperature of the Indian Ocean is increasing faster than other oceans [3]. These factors increase flood risk for coastal zones, particularly those in low lying areas, such as Oman.

These increased risks are expected to persist. For example, in Japan, general circulation models used to forecast the frequency of future storms and cyclones indicate potential losses of about 10 billion USD per year from 100 year return period rainfall events [4]. Similar findings from more frequent and extreme events are projected for central India [5]. Major storms also heavily impact important coastal habitat and ecosystems, particularly coral reefs [6, 7] that are also under climate change pressure from ocean warming and acidification. Recent studies provide a projection of tropical cyclones (TCs) for 2081–2100 at global and regional scales, and conclude that with a scenario of 2 degrees of global warming, there will be a global reduction in the number of TCs but an increase in TC intensity (more cyclonic events of category 4 and 5), and an increase of tropical cyclonic rainfall amount by 5–20% [8].

These more intense events are often very damaging. For example, the Gulf Coast of the United States was hit by Hurricane Katrina in 2005, resulting in 1300 fatalities, and infrastructure and urban damage costs of $75 billion. In 2003, Hurricane Isabel struck North Carolina and New Jersey resulting in 40 deaths and $3.6 billion in damage [9]. The damage risk is exacerbated by urbanization, including in the Gulf countries, as flood hazards are elevated by modification of otherwise natural land surfaces, whilst more people and assets are also exposed to the elevated flood risk [10].

These pressures apply in Oman, where TC's represent a significant risk to people and infrastructure [11]. TCs and storms coming from the Arabian Sea are common in Oman [12, 13]. They are associated with intense rainfall, flash flooding, and can generate tremendous infrastructural, socio-economic and environmental losses [14–16]. For example, TC Gonu, the first category 5 'super cyclone' recorded in the Arabian Sea in a century, hit northern Oman and Iran in 2007, with 78 fatalities, and an estimated $4.6 billion in damage [12]. Gonu is considered Oman's worst natural disaster but smaller events can be very damaging too; for example a 2002 cyclonic storm (ARB 01) caused about $50 million in damage [17, 18]. The most intense recent events to impact Oman were in the 2015 North Indian Ocean season, comprising the cyclonic storm Ashobaa and the extreme cyclonic storms Chapala and Megh, both of which resulted in fatalities. Since then southern Oman has experienced flash flooding due to TCs Mekuno and Luban in 2018, with further intense storm events in 2020 [19]. A higher frequency of extreme rainfall events is now considered the new norm in Oman [20].

In the Arabian Sea region, the majority of TCs form near the Laccadive Islands [~11° N, 73° E] in two seasons: the pre-monsoon and the post-monsoon [21]. The pre-monsoon season runs from the end of April to June when the south-west wind rises, and the sea surface becomes very warm. The post-monsoon season runs from September to December, when the south-west wind declines and a northeast wind develop over the Arabian Sea [22].

Recent studies report an increase in the frequency of extremely severe cyclonic storms in the pre-monsoon period [23] an increase attributed to elevated anthropogenic black carbon and sulphate emissions [24]. These anthropogenic emissions lead to a weakening of the climatological vertical wind shear, causing an increase in TC intensity in the Arabian Sea [24].

Most of these more intense TCs make landfall [24] and so pose a growing risk. Understanding cyclone frequency and direction is thus essential in identifying areas at high risk in Oman and supporting disaster management. Therefore, this study analyses TCs in the Arabian Sea to better understand their frequency and direction. The study presents a spatio-temporal analysis of cyclone tracks in the Arabian Sea region, drawing on observations from 1881–2019. The research aims to identify high-frequency seasons, the cyclone direction in each season, and the linear direction trend. The results are intended to identify the more exposed areas around the Arabian Sea, particularly in Oman, and to support disaster risk appraisal and management.

2. Data and methods

A spatial [GIS] database of TCs was created based on tracks obtained from the Indian Metrological Department Atlas for the period 1881–1999 [25, 26] and the

IMD e-Atlas for storm and depression tracks over the North Indian Ocean for 1891–2019 [27]. The data used represent all known major tropical systems in the Arabian Sea, 1881–2019, for which observations are available. The tracks show tropical system intensity [category] with data organized based on the temporal [seasonal and monthly] distribution of cyclone tracks and their point of origin. Tracks that made landfall in Oman were extracted from the total tracks for dedicated analysis.

Two methods were used in the spatio-temporal analysis. First, kernel density estimation [KDE] was used to calculate the TCs density distribution. The standard deviations of the KDE showed the high-density area of tracks and were calculated seasonally and monthly for both total Arabian Sea cyclones, and those making landfall in Oman.

Next, the Linear Direction Mean [LDM] is used to identify the linear trend of tracks [the mean orientation and mean angular direction of cyclone tracks], also by month and season. Adding the temporal dimension is useful as it can reveal seasonal variability in tracks. The LDM statistic is "the angle of a line representing the mean direction or orientation of all the lines in the dataset" [28]. The orientation mean considers only the tracks' movements, whilst the linear direction mean [θ_R] additionally considers the from/to [e.g. east to west] direction of travel. Esri ArcGIS was used to calculate the LDM and circular statistics of compass angle and circular variance [see below] with tracks defined by each cyclone's origin and endpoint coordinates plotted in a space graph with origin 0, 0 [28]. The statistics calculated are as follows:

$$Y = \frac{\sum_{i=1}^{n} \sin \theta i}{n} \tag{1}$$

$$X = \frac{\sum_{i=1}^{n} \cos \theta i}{n} \tag{2}$$

Where X and Y are the rectangular coordinates of the mean point, and n is the number of tracks.

$$OR = \sqrt{X^2 + Y^2} \tag{3}$$

Where OR is the length of the resultant vector [in decimal degree [DD] and subsequently distance, where 1 DD = 111.3 km].

$$\cos \bar{\theta} = \frac{X}{OR} \tag{4}$$

$$\sin \bar{\theta} = \frac{Y}{OR} \tag{5}$$

Where $cos\,\bar{\theta}$ is the sum of the cosines of the angles of the trackand $sin\,\bar{\theta}$ is the sum of the sines of track angles. The mean angular direction θ_R is then:

$$\boldsymbol{\theta_R} = \arctan\left(\frac{\sin \bar{\theta}}{\cos \bar{\theta}}\right) \tag{6}$$

The θ_R of the resultant vector $[R]$ is the mean directional counter clockwise from due east [with a value up to 180°]. The resultant vector $[R]$ shifted into the correct quadrant is the compass angle clockwise from due north [with value 0–360°].

The circular variance [s] shows how much storm track directions deviate from the directional mean, and is analogous to the standard deviation. The circular variance is calculated from the length of the resultant vector [*OR*], with the result then subtracted from 1 as [28]:

$$s = 1 - \frac{OR}{n} \tag{7}$$

The circular variance, *s*, has a value between 0 and 1. If all lines pointed in the same direction, the *OR* would equal the number of lines n [*OR*/n = 1], and the circular variance [*s*] = 0. If lines pointed in the opposite direction, then *OR* = 0 and *s* =1 [28].

The KDE and LDM analyses thus show the location-frequency [density] of the Arabian Sea tropical cyclones [1881–2019] and their mean tracks, and how these factors vary over time, by cyclone intensity, and in particular for those cyclones which pose the most significant risk to Oman, those making landfall.

3. Results

3.1 Arabian Sea tropical cyclones frequency

Table 1 presents summary statistics for TCs that formed in the Arabian Sea, 1881–2019. In total, 236 systems formed in the Arabian Sea; of which 134 made landfall, and 102 died in the Arabian Sea or the Gulf of Aden. India has the highest frequency of Arabian Sea cyclones making landfall, with 26.7% of the total, about half of all those making landfall. Overall 47 systems made landfall in Oman [19.9% of the total], and another 16 entered Omani coastal waters but died at sea [between 60–64° E].

Table 2 shows the monthly distribution of Arabian Sea cyclones. The data is grouped into temporal bands that indicate two to four cyclones occur every 15 years, except for 1955–1984 when twice this rate occurred. There is a high frequency of cyclone formation in May–June and October–November. Half of all landfall events occurred in the pre-monsoon [13 in May, nine in June], and 30% in the post-monsoon [five in September, six in October, plus one in November], a pattern consistent with the shorter record of Membery [29]. Thus the formation of TCs in the Arabian Sea occurs in two distinct seasons; pre-and post-monsoon.

Landfall country	Frequency	%
India	63	26.7
Oman	47	19.9
Pakistan	10	4.2
Somalia	7	3.0
Yemen [Socotra islands]	7	3.0
Terminates at Sea	102	43.2
Arabian Sea	96	40.7
Gulf of Aden	6	2.5
Total	236	100

Source: [IMD archive].

Table 1.
Distribution of tropical systems in the Arabian Sea, 1881–2019.

Years	Jan	Feb	Mar	Apr	May	June	July	Aug	Sep	Oct	Nov	Dec	Total	%
1880–1894	0	0	0	1	2	5	0	0	0	1	3	0	12	5
1895–1909	0	0	0	1	4	5	2	0	0	5	1	1	19	8
1910–1924	0	0	0	1	3	2	0	0	0	2	1	0	9	6
1925–1939	1	0	0	0	5	7	0	0	2	5	4	0	24	9
1940–1954	0	0	0	4	2	6	0	0	1	4	5	1	23	10
1955–1969	0	0	0	0	6	5	3	2	1	5	5	2	29	13
1970–1984	0	0	0	1	6	9	4	0	2	12	10	2	46	21
1985–1999	0	0	0	0	3	6	0	0	1	6	5	3	24	10
2000–2014	0	0	0	0	5	9	0	0	4	6	7	1	32	12
2015–2019	0	0	1	0	2	4	0	0	1	5	1	4	18	6
Total	1	0	1	8	38	58	9	2	12	51	42	14	236	100
%	0%	0%	0%	3%	38%	25%	4%	1%	5%	22%	18%	6%	100	

Source: [IMD archive].

Table 2.
Monthly distribution of tropical systems in the Arabian Sea, 1881–2019.

3.2 Track analysis of Arabian Sea storms making landfall in Oman

3.2.1 Tracks classification

The tracks analysis (**Figure 1**) reveals a distinct difference in pre-and post-monsoon cyclones. **Figure 1**(1) shows the distribution of cyclone origin point and track in the pre-monsoon by month. All pre-monsoon cyclone origins were in the Arabian Sea, to the south-east in May, moving slightly to the northeast in June, then to the north in July. The tracks vary in each month direction, but a clear pattern of track movement is to the south-west Arabian Sea in May and to the north-west in June and July, although there are numerous cases in June when the track curves northeast toward India.

Figure 1(2) shows the distribution of origin points and tracks in the post-monsoon. Origin points are distributed over a larger area than in the pre-monsoon. Most post-monsoon storms originate in the northeast Arabian Sea in September and move gradually to the south-east and the south Arabian Sea in October–December. However, several storms formed in the Bay of Bengal and track west over India before arriving in the Arabian Sea. The tracks analysis shows that in the post-monsoon, cyclones usually track to the west of the Arabian Sea in September, then gradually to the south-west toward the Gulf of Aden and the Horn of Africa in October–December. However, some November cyclones recurved to India in the east of the Arabian Sea.

The origins and tracks for the sub-set of systems that made landfall in Oman also vary by season and month of formation. **Figure 1**(1) shows pre-monsoon origins and tracks, revealing a distinct difference in tracks within this period. In May, systems frequently travel to the coast between Masirah Island (central Oman) and Salalah in south-west Oman. In June the track direction moves to central to north-east Oman, such that tropical systems arrive at the coast between Masirah Island and Ras Al Had, the easterly most point of Oman at the mouth of the Oman Sea. Note, however, that the historical record indicates that storms occasionally deviate from this general pattern. A strong storm of May 1898 crossed Oman from Ras

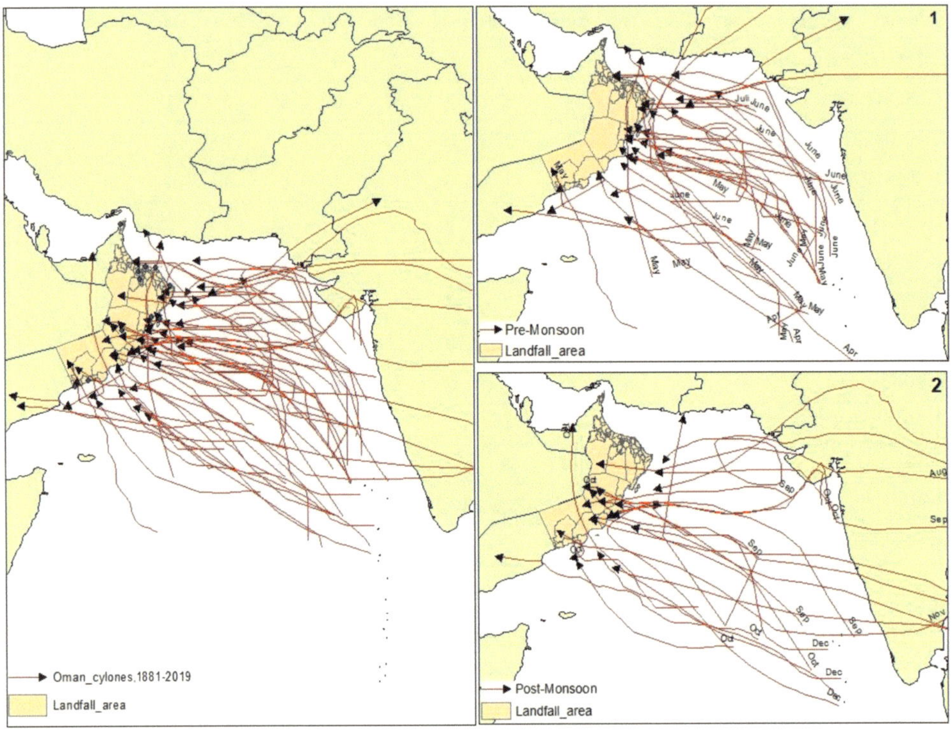

Figure 1.
Seasonal distribution of Arabian Sea cyclone tracks with landfall in Oman, 1881–2019: (1) pre-monsoon, (2) post-monsoon.

Madrakah (south of Masirah Island) and moved to north Oman [25], whilst a storm of June 1885 moved to the south-east coast and entered the Gulf of Aden off Yemen [30].

Figure 1(2) shows that the origins and tracks in the post-monsoon similarly have a distinct tracks pattern within this period. In September cyclones track to Central Oman from Masirah Island to Ras Madrakah and then in October–December move progressively toward Salalah in south-East Oman.

3.2.2 KDE of Arabian Sea tracks

Kernel Density Estimation [KDE] is used to analyze cyclone tracks' distribution in the Arabian Sea and identify areas with a high-density of cyclones. **Figure 2** displays the KDE of all tracks in the Arabian Sea and shows a high density of tracks over a large area of the Arabian Sea, from 15–25° N and 60–73° E. In the pre-monsoon (**Figure 2**(1)) there is a high-density of track movement to the north-east, toward the north-west Indian coast (Gujarat), whilst in the post-monsoon (**Figure 2**(2)) the high-density area is more to the south and south-east of the Arabian Sea.

Figure 3 shows the KDE of those tracks that made landfall in Oman. There is a high density of tracks in the mid-East coastline of Oman near Ras Madrakah, which is also evident in the pre-monsoon period (**Figure 3**(1)). In the post-monsoon period, the highest density of tracks is toward the south-east coastline of Oman, near Salalah (**Figure 3**(2)).

Figure 4 shows the KDE of cyclone tracks making landfall in Oman, by pre-and post-monsoon month. In May, the tracks' highest density is in mid-Oman near Ras Madrakah, with some tracks in the south near Salalah. In June there remains a high

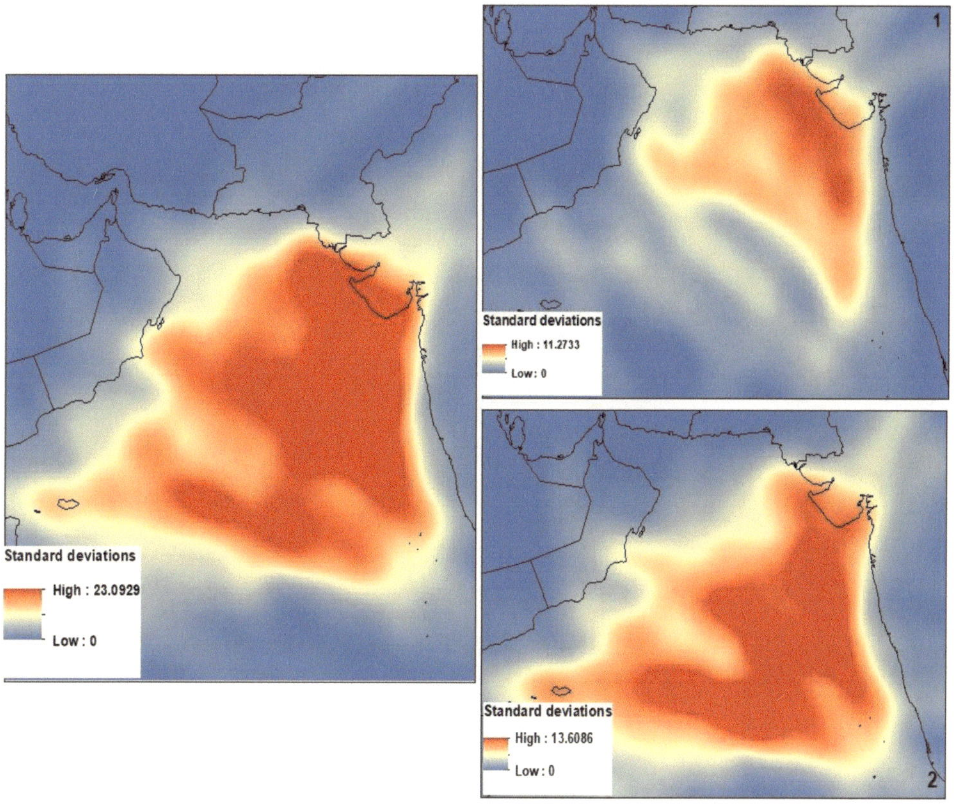

Figure 2.
Kernel density estimation (KDE) of Arabian Sea tracks 1881–2019: (1) pre-monsoon, (2) post-monsoon.

density in mid-Oman, but additionally to the north of Oman, near Ras Al Had. In the post-monsoon, the highest density of the tracks is in mid-Oman near Ras Madrakah in September which moves to the south near Salalah in October.

3.2.3 Linear direction mean of Arabian Sea storm tracks

Linear Direction Mean (LDM) results are presented in **Table 3** and **Figure 5**. For all storm tracks in the Arabian Sea (**Figure 4**(1)) the LDM is to the north-west toward Oman and Iran, with directional mean angle 127.8° (clockwise from due east; compass angle 322.25°), and an average track length of 1480 km. **Figure 4**(1) shows that the LDM in the pre-monsoon is to the north, toward Pakistan, with mean directional angle 120.4°, the post-monsoon LDM is to the north-west toward Oman's north-east coastline, with mean directional angle 136.4°.

Figure 5(1) shows the LDM of those tracks that made landfall in Oman. This track's mean directional angle is 157.8°, to the mid-East Oman coastline south of Masirah Island, and with a mean length of 2169 km. The LDM of the pre-monsoon tracks moved slightly to the north of the LDM of all tracks, to the middle of Masirah Island, with mean directional angle 146.0°, and mean length 1827 km. The post-monsoon landfall tracks have a mean directional angle of 157.9° to Ras Madrakah, with a mean length of 2361 km.

Figure 5(3) shows the LDM of tracks that made landfall in Oman, by seasonal distribution (pre-and post-monsoon), and **Figure 5**(4) shows the monthly LDM of tracks. May and June in the pre-monsoon are the highest frequency storm months.

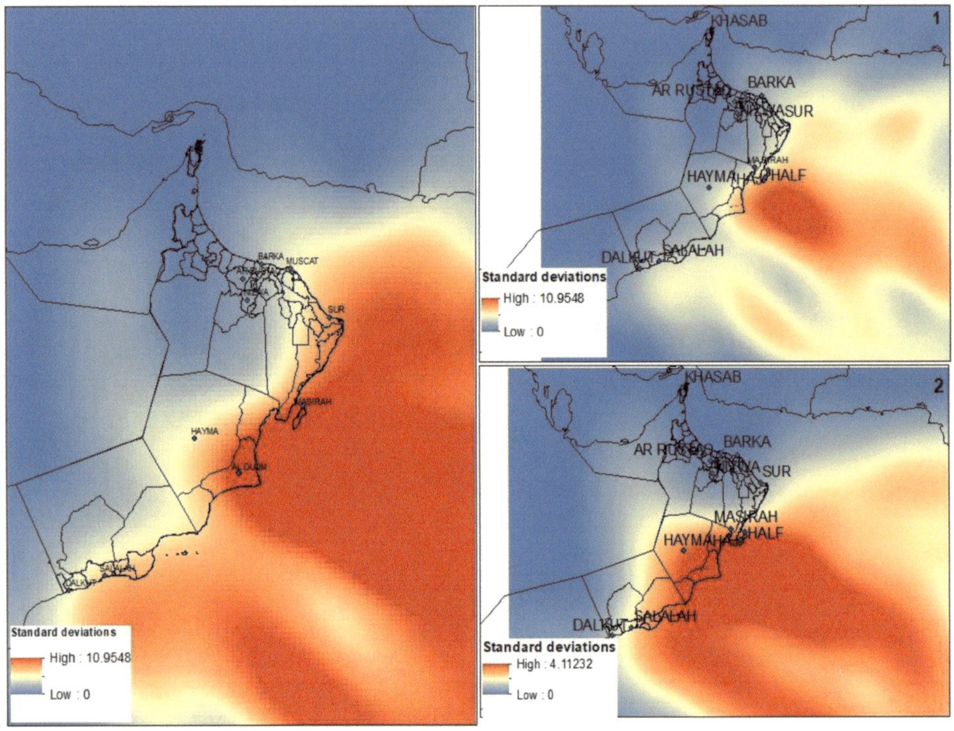

Figure 3.
Seasonal kernel density estimation [KDE] of Arabian Sea cyclone tracks 1881–2019, for cyclones making landfall in Oman. (1) pre-monsoon, (2) post-monsoon.

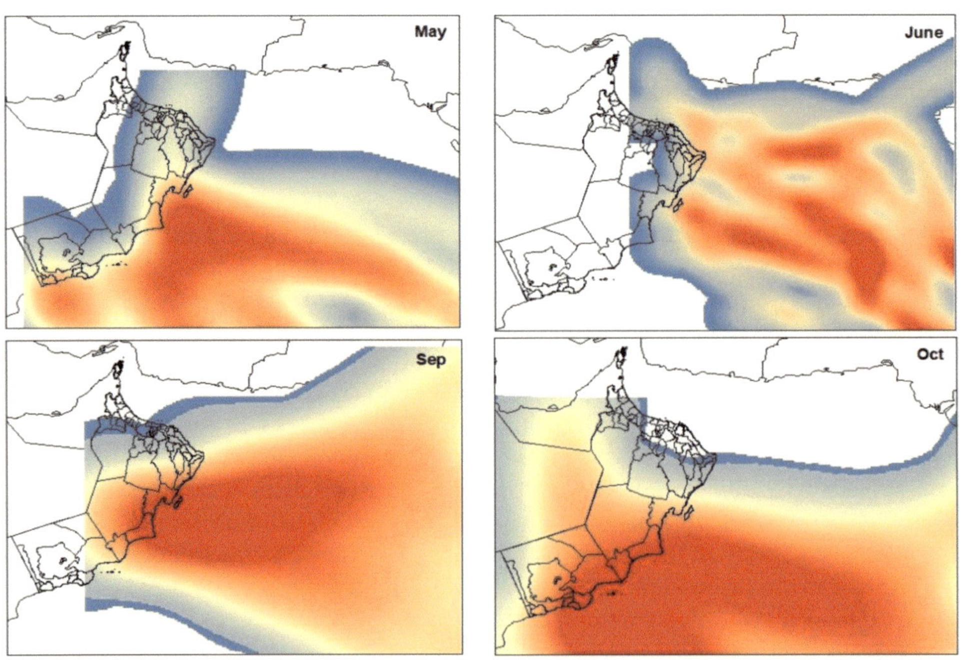

Figure 4.
Monthly kernel density estimation [KDE] of Arabian Sea cyclone tracks 1881–2019 for cyclones making landfall in Oman.

Tracks	Compass Angle [1]	Direction Mean angle[2] $[\theta_R]$	Circular Variance $[s]$	X	Y	*OR* [km]
All tracks	322.25	127.75	0.28	67.72	16.82	1479
Pre-monsoon	332.59	120.41	0.24	67.04	18.96	1221
Post-monsoon	313.63	136.37	0.31	66.27	14.91	1569
Tracks making Oman landfall	298.25	151.75	0.11	64.54	17.09	2169
Pre-Monsoon	302.25	147.74	0.08	63.47	17.11	1827
Post-Monsoon	287.39	162.60	0.03	64.71	16.25	2361
May	302.59	147.41	0.05	62.30	15.71	1748
June	306.30	143.70	0.12	65.35	18.91	1984
Sep	287.63	162.37	0.02	67.99	18.92	2424
Oct	289.22	160.78	0.03	65.87	15.45	2885

Notes. 1. The resultant vector [R] shifted into the correct quadrant clockwise from due north. 2. The directional mean angle $[\theta_R]$ of the resultant vector [R] value up to 180° counter clockwise from due east.

Table 3.
Linear direction mean (LDM) parameters for Arabian Sea storm tracks, 1881–2019.

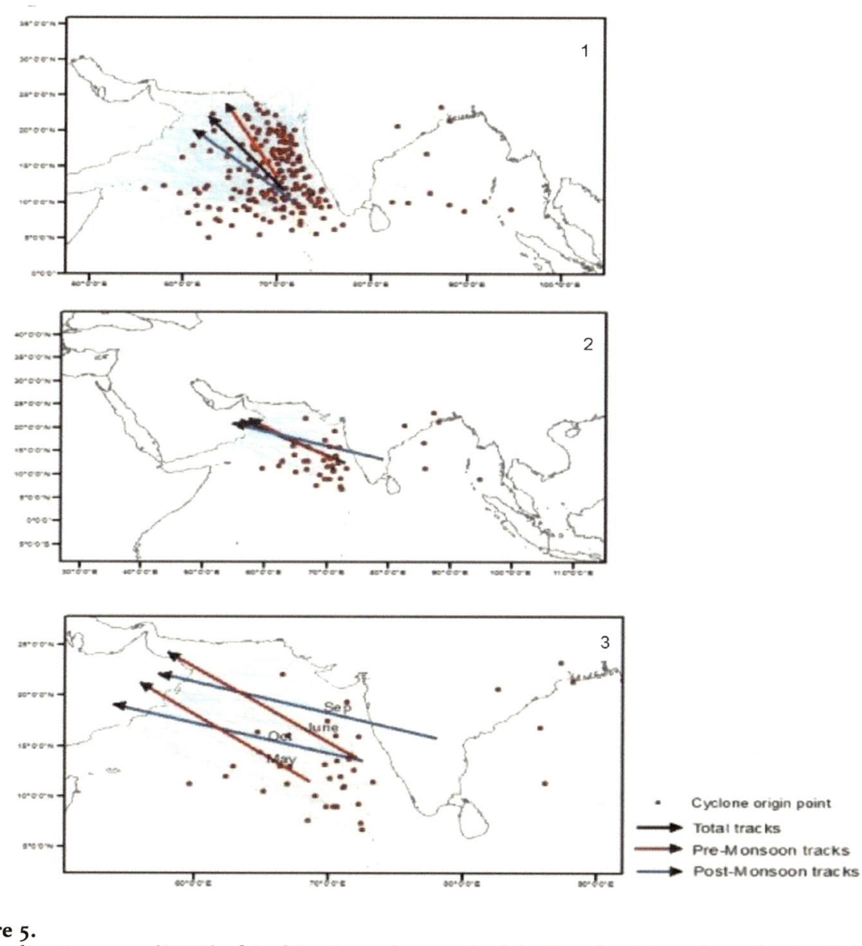

Figure 5.
Linear direction mean (LDM) of Arabian Sea cyclone tracks: (1) all tracks, (2) tracks making landfall in Oman, and (3) tracks making landfall in Oman by pre- and post-monsoon month.

Storms in May track toward the mid-east coast of Oman and Ras Madrakah. In June storms have a similar LDM but move to the north-west Arabian Sea and toward Oman's capital, Muscat, as they originate further north and east than May storms, whilst having a track length several hundred kilometers greater. A similar situation occurs in the post-monsoon. The LDM of tracks is similar for September and October storms, with October storms originating to the northeast of May storms. September storms thus tend to track toward Masirah Island on Oman's eastern coast, whilst in October, storms making landfall tend to head toward the south-east coast, and Salalah, Oman's main southern city.

4. Discussion

Our study has sought to support the management of risk from natural hazard in Oman, through developing an improved understanding of TC hazard. To do this we have: collated geographical data from the Indian Metrological Department for all known TC and cyclonic storm events in the Arabian Sea since 1881; plotted the tracks of each storm from point of origin to expiration, by storm category (size), by month and season, and with particular reference to those events that made landfall in Oman (these events tend to be large); and conducted a directional analysis to determine likely course of a TC and a spatial (KDE) analysis to determine geographic areas at higher risk of TC makings landfall.

Our analysis sheds light on the origins and tracks of Arabian Sea TCs by season. TCs in the Arabian Sea form in two seasons: the pre-monsoon and the post-monsoon [30, 31]; our analysis reveals a distinct difference in the subsequent TC track between these pre-and post-monsoon periods. All pre-monsoon TCs form in the Arabian Sea itself. In May, they start in the south-east and track to the south-west; in June their origin moves north-east and TCs track north-west; in July TCs form further north still, but continue to track north-west. There are however cases in June when TCs curve north-east toward India. TCs in the post-monsoon period originate over a larger area extending as far as the Bay of Bengal, with TCs tracking west over India to arrive in the Arabian Sea. However, most TCS in this season form in the north-east Arabian Sea in September, then south-east and the south Arabian Sea from October – December. TCs tend to track south-west toward the Gulf of Aden and the Horn of Africa in October–December, although some TCs in the east Arabian Sea recurved to India.

In general, TCs affect Omani coastal areas from Muscat in the north to Salalah in the south (**Figure 6**). Our analysis specifically of those TCs that made landfall in Oman shows that the origins and tracks for systems that make landfall in Oman also vary by season and month of formation. Pre-monsoon TCs tend to originate in the south of the region in May and frequently make landfall between Masirah Island (central Oman) and Salalah in south-west Oman. In June TCs tend to originate further north and make landfall between Masirah Island and Ras Al Had, the easterly most point of Oman at the mouth of the Oman Sea. There are however, some exceptions to this general behavior. Post-monsoon cyclones also have distinct origins and tracks. September TCs tend to form in the north and are likely to make landfall in central Oman, from Masirah Island to Ras Madrakah (Al Duqum), whilst from October–December TCs move progressively toward Salalah in south-east Oman.

Our tracks analysis (part of a wider study of TC resilience in Oman) was first conducted up to and including 2014, and subsequent events have provided an opportunity to test our general conclusions. Early confirmation came from tropical storm Ashobaa in June 2015, which our results suggested would strike Oman in the

Figure 6.
Oman's coastal areas.

northern region, and indeed this is what happened, with the storm traveling to the north-eastern coast, hitting Oman near Ras-Al Had. Similarly, TC Mukuno that also developed in the pre-monsoon (May 2018), traveled to the south-eastern coast, hitting Salalah and Yemen, as suggested by our analysis. For the post-monsoon we note that TC Hika, which formed in September 2019, traveled to the mid-Omani coastline toward Al Duqum (where it caused substantial damage), whilst TC Luban, formed in the Arabian Sea in October 2018, took a direction to the south-west Arabian Sea toward Salalah and Yemen. Whilst such observations cannot qualify as a formal test of the analysis, they do provide support for the general conclusions as to the seasonal behavior of cyclone tracks in the region.

Thus our analysis reveals a series of general and broadly predictable spatio-temporal patterns. Individual events may deviate from these trends, but the general patterns are useful in informing natural hazard risk assessment and management in the region, including Oman, which has suffered extensive damage in the past due to TCs. The results could, for example, assist with more targeted cyclone preparation and deployment of emergency response resources, based upon areas most at risk to cyclones overall (strategic planning), and to specific storm events when these are first identified (tactical planning). For example, knowing that cyclones that develop in June are more likely to make landfall in Oman's northern part is valuable emergency planning intelligence.

Globally, the urban area exposed to both flood and drought is forecast to at least double from 2000–2030 [32], indicating the growing importance of understanding and managing these risks. For Oman, the key risk is loss of life and damage from TC. Developing an improved understanding of TC occurrence and behavior, through historical analysis of TC timing and tracks should help to build resilience to TC, and climate change, for the country. Building resilience to TC requires a much wider process of institutional organization and action, which we discuss elsewhere [33] with specific reference to Oman's natural hazard management governance. However, a key element of natural hazard risk management is an understanding of the hazard itself. Our historical analysis of nearly a century and a half of TC tracks in the Arabian Sea, provides enhanced insight into the likely timing and location of TCs making landfall in Oman. This knowledge can support strategic decision making (e.g. prior appraisal of potential flood damage; location prioritization of mitigation resources) and improved understanding of likely destination of TC when they emerge in the Arabian Sea, thus supporting pre-emptive event management.

5. Conclusions

In Oman, tropical cyclones and storms are amongst the most significant natural hazards currently facing the country. Past TCs have resulted in significant loss of life, and billions of dollars in damage costs. The future incidence and behavior of TCs is uncertain, but prediction of a global reduction in TC frequency, but accompanied by an increase in intensity and rainfall when TCs do occur [8], seems to be borne out by recent decades of TC observation in the Arabian Sea region, where fewer but more extreme events are evident.

From our analysis of all recorded Arabian Sea cyclones, we conclude that cyclone tracks vary according to season (pre-and post-monsoon) and month of formation. In the pre-monsoon, cyclones tend to form in the northeast Arabian Sea and moves to the north; in the post-monsoon cyclones tend to form to the southeast Arabian Sea and track to the west.

Whilst there are clearly exceptions evident in the historical record of Arabian TCs spatio-temporal analysis reveals clear seasonal and monthly patterns in the origins and tracks of TCs. This leads to an improved estimation of the likely fate of tropical systems, including a better understanding of where TC s are likely to make landfall in Oman, given their time and point of origin. This knowledge can help improve disaster planning for areas at high-risk.

Author details

Suad Al-Manji[1]*, Gordon Mitchell[2] and Amna Al Ruheili[3]

1 Ministry of Education, Muscat, Oman,

2 University of Leeds, Leeds, UK

3 Sultan Qaboos University, Muscat, Oman

*Address all correspondence to: suad.almanji@gmail.com

References

[1] Jonkman, S. N. Global perspectives on loss of human life caused by floods. Natural Hazards. 2005; 34[2], 151–175

[2] Herath, S. M., and Sarukkalige, R. Evaluation of empirical relationships between extreme rainfall and daily maximum temperature in Australia. Journal of Hydrology. 2018; 556, 1171–1181

[3] Roxy, M. K., Ritika, K., Terray, P. and Masson, S. The curious case of Indian Ocean warming. Journal of Climate. 2018; 27[22], 8501–8509

[4] Kazama, S., Sato, A., and Kawagoe, S. Evaluating the cost of flood damage based on changes in extreme rainfall in Japan. Sustainability Science. 2009; 4 [1], 61

[5] Naidu, C. V., Raju, A. D., Satyanarayana, G. C., Kumar, P. V., Chiranjeevi, G., and Suchitra, P. An observational evidence of decrease in Indian summer monsoon rainfall in the recent three decades of global warming era. Global and Planetary Change. 2015; 127, 91–102

[6] Knutson, T. R., McBride, J. L., Chan, J., Emanuel, K., Holland, G., Landsea, C. et al. Tropical cyclones and climate change. Nature Geoscience. 2010; 3[3], 157–163

[7] Cheal, A. J., MacNeil, M. A., Emslie, M. J., and Sweatman, H. The threat to coral reefs from more intense cyclones under climate change. Global Change Biology. 2017; 23[4], 1511–1524

[8] Walsh, K. J., McBride, J. L., Klotzbach, P. J., Balachandran, S., Camargo, S. J., Holland, G. and Sugi, M. Tropical cyclones and climate change. Wiley Interdisciplinary Reviews: Climate Change. 2016; 7[1], 65–89

[9] Bourque, L. B., Siegel, J. M., Kano, M., and Wood, M. M. Weathering the storm: The impact of hurricanes on physical and mental health. The Annals of the American Academy of Political and Social Science. 2006; 604[1], 129–151

[10] Scholz, W., Kaiser, M., and Pallasch, M. The Potentials and Risks of Wadis in Cities in the Gulf Region. In Handbook of Quality of Life and Sustainability. 2020; 497–520. Springer, Cham.

[11] El-Saadawy, O., Gaber, A., Othman, A., Abotalib, A. Z., El Bastawesy, M., and Attwa, M. Modeling Flash Floods and Induced Recharge into Alluvial Aquifers Using Multi-Temporal Remote Sensing and Electrical Resistivity Imaging. Sustainability. 2020; 12[23], 10204

[12] Evan, A. T., and Camargo, S. J. A climatology of Arabian Sea cyclonic storms. Journal of Climate. 2011; 24[1], 140–158

[13] Byju, P., and Kumar, S. P. Physical and biological response of the Arabian Sea to tropical cyclone Phyan and its implications. Marine Environmental Research. 2011; 71[5], 325–330.

[14] Al-Rawas, G. A., and Valeo, C. Relationship between wadi drainage characteristics and peak-flood flows in arid northern Oman. Hydrological Sciences Journal–Journal des Sciences Hydrologiques. 2010; 55[3], 377–393

[15] Al-Qurashi, A., McIntyre, N., Wheater, H., and Unkrich, C. Application of the Kineros-2 rainfall–runoff model to an arid catchment in Oman. Journal of Hydrology. 2008; 355 [1–4], 91–105

[16] Momani, N. M., and Fadil, A. S. Changing public policy due to Saudi City of Jeddah flood disaster. Journal of Social Sciences. 2010; 6[3], 424–428.

[17] Al Ruheili, A., and Radke, J. Visualisation of 2002 storm surge along the coast of Dhofar, case study of Oman. Environment, Development and Sustainability. 2018; 22[1], 501–517

[18] Al Ruheili, A., Dahm, R., and Radke, J. Wadi flood impact assessment of the 2002 cyclonic storm in Dhofar, Oman under present and future sea level conditions. Journal of Arid Environments. 2018; 165, 73–80

[19] Al Ruheili,A. The Aftermath of Mekuno at Jarziz Watershed in Dhofar. Modern Environmental Science and Engineering. 2020; 6[4], 494–499

[20] Gunawardhana, L. N., Al-Rawas, G. A., and Al-Hadhrami, G. Quantification of the changes in intensity and frequency of hourly extreme rainfall attributed climate change in Oman. Natural Hazards. 2018; 92[3], 1649–1664

[21] Galvin, J. The weather and climate of the tropics: Part 7–Tropical revolving storms, Weather. 2008; 63[11], 327–333

[22] Galvin, J. The weather and climate of the tropics: Part 6–Monsoons, Weather. 2008; 63[5], 129–137

[23] Murakami, H., Vecchi, G. A., and Underwood, S. Increasing frequency of extremely severe cyclonic storms over the Arabian Sea. Nature Climate Change. 2017; 7[12], 885

[24] Evan, A. T., Kossin, J. P., and Ramanathan, V. Arabian Sea tropical cyclones intensified by emissions of black carbon and other aerosols. Nature. 2011; 479 [7371], 94

[25] Indian Metrological Department. Atlas of the Tracks of storms and depressions in the Bay of Bengal and the Arabian Sea 1877–1970. [Internet] 1979. http://www.imdchennai.gov.in/cyclone_eatlas.htm. [Accessed 1st Aug 2020]

[26] Indian Metrological Department. Atlas of the Tracks of storms and depressions in the Bay of Bengal and the Arabian Sea 1971–1995. [Internet] 1999. http://www.imdchennai.gov.in/cyclone_eatlas.htm. [Accessed 1st Aug 2020]

[27] Indian Metrological Department. Tracks of storms and depressions in the Bay of Bengal and the Arabian Sea 1971–1995. [Internet] 2019. http://www.imdchennai.gov.in/cyclone_eatlas.htm. [Accessed 1st Aug 2020]

[28] Mitchell, A. The Esri guide to GIS analysis, volume 2: Spatial measurements and statistics, 1st edn. 2005. Redlands, CA: Esri Press

[29] Membery, D. Famous for 15 minutes: An investigation into the causes and effects of the tropical storm that struck southern Arabia in June 1996, Weather. 1998; 53[4], 102–110

[30] Membery, D. Monsoon tropical cyclones: Part 2, Weather. 2002; 57[7], 246–255

[31] Membery, D. Monsoon tropical cyclones: Part 1, Weather. 2001; 56[12], 431–438

[32] Güneralp, B., Güneralp, İ., and Liu, Y. Changing global patterns of urban exposure to flood and drought hazards. Global Environmental Change. 2015; 31, 217–225

[33] Al-Manji, S, Lovett, J. and Mitchell, G. Factors affecting disaster resilience in Oman: Integrating Stakeholder Analysis and Fuzzy Cognitive Mapping. Risk, Hazards and Crisis in Public Policy. 2021; 12, 1, doi.org/10.1002/rhc3.12201

Chapter 11

Effects of Demographic Characteristics for Farmers' to Climate Change in Bunkure, Nigeria

Salisu Lawal Halliru, Abubakar Abdullahi Bichi and Aliyu Shu'aibu Muhammad

Abstract

This study examine the effects of demographic characteristics on farmers' level of awareness to climate change in Bunkure, Nigeria. The study hypothesized that, there is no significant relationship between farmers' demographic characteristics and level of awareness to climate change. Survey design approach was adopted using primary data source. A total of three hundred and fourty-seven (347) farmers were selected purposively. The data collected was analyzed using descriptive statistic and ANOVA. The study revealed that majority of the farmers in the area are males that fall within the active age bracket due. However, the study concluded that the greater the size of the household the greater the chances of being engaged in farming and more aware on climate change effects. It was revealed that those that attend tertiary education do not participate in farming activities. Also large number of the respondents were aware that climate is changing through the educative programmes of mass media. The study recommended that, adequate support should be given to female to participate in all agricultural activities. Educated persons should be encourage to participate more on agriculture. More emphasis on dissemination of climate change issues on adaptation strategies through the mass media is also highly needed.

Keywords: demographic, farmers, climate change

1. Introduction

The issue of climate change today has become one of the major challenges that humanity faces in the 21st century due to its significant effects on socio-economic and political stability in different geographical location. The Intergovernmental Panel on Climate Change (IPCC) [1], views climate change as statistically significant variation in either mean state of the climate or in its variability, persisting for an extended period (typically decades or longer). This change may be due to natural processes or persistent anthropogenic activities over the different component of the ecosystem. Although extreme violent weather has occurred throughout the human history, but the recent upsurge in climate related hazards call for the confirmation of the argument for global warming and climate change as results of observed

IntechOpen

increasing in temperature that makes some areas to experiencing extreme weather conditions [2–7].

The on-going climate change and its associated global warming are expected to cause distinctive climate variability in different zones, which will impact negatively on the ecosystem and agriculture [8, 9]. That is the reason why, it is good for the people to acquire knowledge of weather and climate and their implications on environmental resources including agriculture for the betterment ecosystem and humanity. In fact the impact of climate change is felt worldwide with varying degree as a result of differences in underlying factors that triggered the change. In Nigeria for instance, the severity of the impact is felt by the majority of people whose livelihoods are more defended on agriculture due to its highly depended on weather and climate. Also agriculture in Nigeria is a major user of land resources as it account for about 1.4 billion hectares of land for crop cultivation and an additional 2.5 billion hectares are used for pasture [10]. With this development, one can say that Climate plays an important role in the living and livelihood of man because his agricultural activities are so much defended on the nature and variability of the Climate.

As Man begin to witness reduction in agricultural production he assumed to be exacerbated by climate change and related events [11]. This scenario makes local farmers to be no longer able to predict onset and cessation of rain, based on past experiences due to their low level of socio-economic status. These lead local farmers with low socio-economic status to plant too early without receiving update on the weather forecast. Hence, it was believed that wealthier farmers benefit more from the daily, week, monthly and seasonal information of weather and climate than the poor subsistence farmers who are the most vulnerable most to challenges of climate change. This problem and other related call the attention of the United Nations Task Team on Social Dimensions of Climate Change to discuss and ensure that the social dimensions of climate change are adequately reflected in global agendas in order to builds on the principles of equity and social justice, especially for the most vulnerable people.

Several attempts have been made by scientists to study the effect of climate change on agricultural productivity and farmers' adaptation in different parts of Nigeria [12]. But information on the effects of farmers' demographic characteristics as determinant factor influencing their level of awareness on climate change in Kano state and Bunkure Local government Area has been limited.

Bunkure local government area is an agrarian rural areas of Kano where most of its inhabitant depend on farming as their main source of livelihood. This trend makes them to be most prone to the effects of climate change and its variability. Therefore is there need to understand the effects of social dimension of the farmers' awareness on climate change and their sources of information to climate change. Understanding this knowledge may greatly help in developing measures and effective monitoring, adaptation and mitigation measures to climate change in the study area and any climate harzard prone areas.

It is against this background that this study was conducted to assess the effects of socio-economic characteristics on farmers' level of awareness to climate change in Bunkure Local Government area of Kano State, Nigeria that will help to mitigate the effects of climate change.

2. Material and methods

2.1 The study area

Bunkure Local Government Area is about 50 km south from the ancient city of Kano located between Latitude $11^{0}34'$ 02"N to Latitude $11^{0}46'$ 05"N of the Equator and between Longitude $8^{0}26'\ 36''$E to Longitude $8^{0}46'\ 43''$E of the Prime Meridian. The study area is comprises of fifteen wards [13] with an aerial extend of 487Km^2 with a population of 170,891 inhabitants [14]. It bordered with Dawakin kudu and Kura LGAs by the North, Wudil and Garko LGA by the East while Kibiya at the South Western part of the study area (see **Figure 1**). The climate of the study area is the reincarnation of the climate of Kano which is the tropical dry-and-wet with the movement of the Inter-Tropical Discontinuity (ITD) gives rise not only the two seasons (wet and dry seasons) but four distinct seasons, *Rani, Damina, Kaka* and *Bazara* [13, 15].

The wet season lasts from May to mid-October with a peak in August while the dry season extends from mid-October of one calendar-year to mid-May of the next. The mean annual rainfall is between 800 mm and 900 mm; and variations about the mean annual values are up to ±30%. The mean annual temperature is about 26 °C [15].

2.2 Methodology

This study adopts survey design approach using primary data in the area in order to assess the effects of socio-economic characteristics on farmers' level of awareness to climate change which in the end will help to mitigate the effects of climate change. The primary data were obtained through questionnaires. The data obtained include information on farmers' socio-economic characteristics, knowledge about climate change and sources of climate change information.

Based on the population figure of the study area, a total of three hundred and eighty-two (382) farmers were selected using the Krejcie and Morgan's [16] method of determining sample size. But after retrieval of the questionnaires only three hundred and fourty-seven (347) was retrieved and the analysis was based on the three hundred and fourty-seven (347). Purposive sampling technique was used for the purpose of selecting respondents. This method is characterized by the use of personal judgment and a deliberate attempt to obtain representative sample

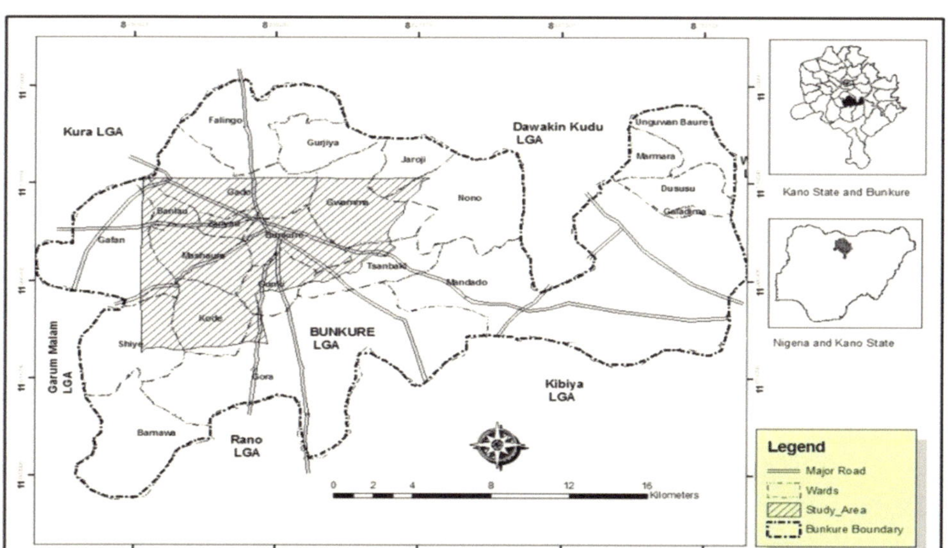

Source: Adopted from Min. of Lands and Planning, Kano State. Redrawn at the Dept. of Geog. BUK (2013)

Figure 1.
Map of the study area.

by including presumable typical areas or groups in the sample [17]. Due to non-availability of population figures at ward level in the 1991 and 2006 census results, the copies of questionnaire were distributed uniformly among the ten [10] wards in the local government area. Each ward had 38 copies of questionnaire except for Bunkure ward which had an addition of 2 making 40 copies of questionnaire being the most populated ward as the local government headquarters.

The data collected was subjected to simple descriptive statistics and tables, percentages and bar graphs was used to present the data in order to analyze the effects of socio-economic characteristics of respondents in relation to their awareness to climate change.

3. Results and discussion

3.1 Socioeconomic characteristics of respondents

3.1.1 Age of respondents

As indicated in **Table 1** majority (79.3%) of the sampled farmers are within the active age bracket of 31–50 years. This results suggest that, the farmers were still in their active/productive age and as such they can understand and experience the effects of climate change and adaptive strategies. The modal age bracket or group of farmers between 21–50 years according to [18] are considered to be active/productive age in farming activities. Participation of youth in agriculture in Nigeria is due to the high level of agricultural apathy by the youth as suggested in the studies on youths' participation in agriculture in Nigeria conducted by [19, 20]. This result corroborated the findings of [21] on adaptation strategies to climate change among grain farmers in Goronyo LGA of Sokoto State which showed that people within the age bracket of 31–50 years are active in farming activities in the area.

3.1.2 Gender of respondents

Gender issues in climate change have recently become important because of the social, spatial and economic contexts within which the change is perceived and responded by gender. The gender distribution of the respondents as shown in **Figure 2** indicated that 95% of the respondents are males while, female's farmers constituted only 5%. The result shows that majority of the farmers are males in

Age Group (Years)	Frequency	Percentage
18–30	42	12.10
31–40	120	34.6
41–50	155	44.7
51–60	20	5.8
>60	10	2.8
Total	**347**	**100**
Source: Field Survey (2019)		

Table 1.
Age of respondents.

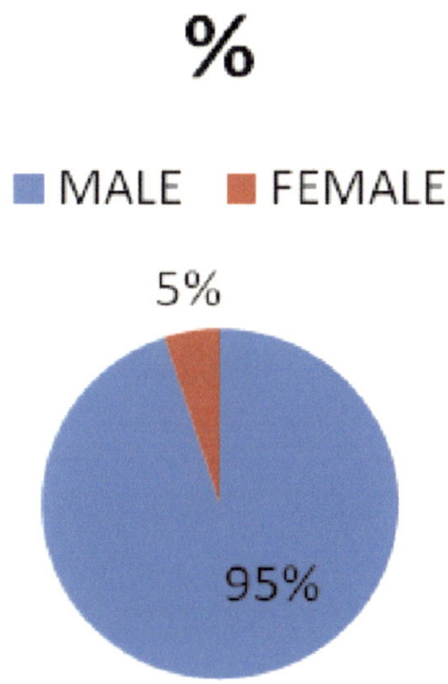

Figure 2.
Distribution of Respondent's gender. Source: Field survey (2019).

the area and this has connection with the culture of the people in Kano where male engaged in all activities including agriculture in order to provide food for their families. This is in agreement with other related studies by [22, 23], that the agricultural sector and the tedious activities related to climate change adaptation strategies are dominated by males. On the other hand, the study contradict the findings of [24] on farmer's perception on the effect of climate change and variation on urban agriculture in Ibadan Metropolis, South-western Nigeria were 66.9% of their respondents were females while 33.1% were males.

3.1.3 Household size

The size of household as indicated in the **Table 2** shows that less than half (45%) of the respondents had a household size of between 6–10 persons while households with 11–15 persons constituted (33%). This implies that on the average, a typical household size is about 11 persons. This indicated that the greater the size of the household the greater the chances of being engaged in farming and more awareness on climate change effects.

Household size	Frequency	Percentage
<5	13	3.8
6–10	157	45.2
11–15	113	32.6
16–20	44	12.7
>20	20	5.8
Total	**347**	**100**
Source: Field Survey (2019)		

Table 2.
Household size.

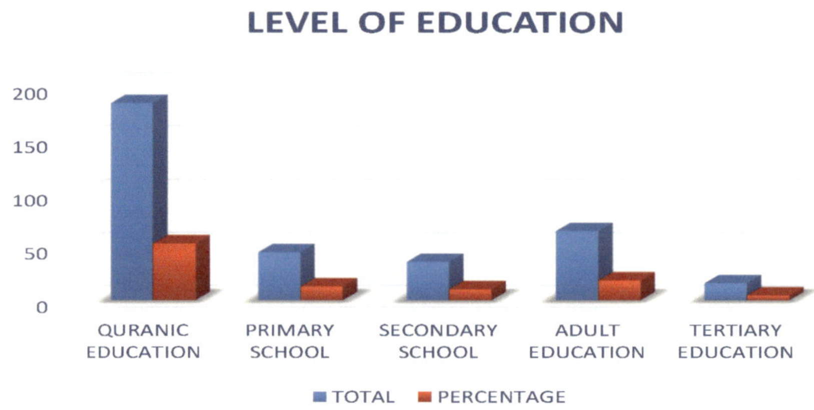

Figure 3.
Level of education of the respondents. Source: Field survey (2019).

3.1.4 Educational attainment of respondent

The results in **Figure 3** revealed that majority of the total respondents (53.31%) had Quranic education as their highest education attainment while a few (4.61%) had tertiary education. In the study conducted by [18] on the Perceptions on climate change and adaptation strategies among sweet potato farming households in Kwara State, North central Nigeria found that just about 2% of the respondents had tertiary education. This indicated that those that attend tertiary education do not participate in farming activities due to the preference for white collar jobs, especially in developing countries like Nigeria [20, 25]. The result in **Figure 3** further indicated that the entire sampled respondents in the study area had one type of education or the other which could enhances their reasoning in all aspect of life.

3.2 Farmers awareness of climate change

3.2.1 Awareness of climate change

From the findings of the study it was revealed in **Table 3** that a large number of the respondents (82%) were aware that climate is changing. Though, the result is higher but lower than the findings of [21] where 98% of the sampled farmers claimed that they are aware of climate change. This difference occurred as a result of differences of their level of educational attainment, culture and socio-economic characteristics. This is because according to [26] in their studies on determinants of farmers' adaptation to climate change in Ghana showed that education was positive and significantly related to farmers' decision to adapt to climate change.

Level of Awareness	Total	Percentage (%)
Yes	284.54	82
No	62.46	18
Total	347	100

Source: Field Survey (2019).

Table 3.
Level of awareness of climate change.

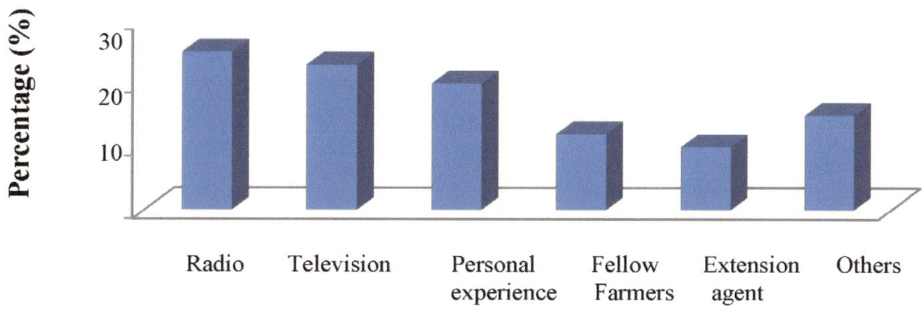

Figure 4.
Source of awareness on climate change. Source: Field survey (2019).

3.2.2 Sources of information on climate change

The results in **Figure 4** shows that farmers access information on climate change from different sources. The figure revealed that mass media were the major sources of awareness for majority of the respondents as 24.5% got their awareness on climate change from the radio; 22.8% from the television. A little over 20% of the respondents claimed that, their awareness of climate change was by their personal experiences. Also extension agents in the study area shows little concern on informing farmers on the incidence of climate change but rather the media plays a significant role in educating not only farmers but the general public through their programmes of daily weather forecast report and other scientific news and programmes. Because today, mass media plays vital role in translating sciences that shaped peoples' perceptions on various issues related to environment, technology and risk . This makes many people to pay much attention to the science based news which in turn affect their positive knowledge about climate change [27].

4. Conclusion

The study provides an insight on the effects of farmers' demographic characteristics to climate change in Bunkure, Nigeria. The study revealed that majority of the sampled farmers in the study area are males that fall within the active age bracket and can understand and experience the effects of climate change due to their active participation in agricultural activities.

As gender issues in climate change become important issues globally the result of the study shows that majority of the farmers are males in the area and this has connection with the culture of the people in Kano where male engaged in all activities including agriculture in order to provide food for their families.

However, the study concluded that the greater the size of the household the greater the chances of being engaged in farming and more aware on climate change effects.

From the results on the level of farmers' education it was revealed that those that attend tertiary education do not participate in farming activities due to the preference for white collar jobs. Also it was established that all the respondents in the study area had one type of education or the other which could enhances their reasoning in all aspect of life.

From the findings of the study it was revealed that a large number of the respondents were aware that climate is changing. Mass media such as radio, television through their programmes of daily weather forecast report and other scientific news and programmes are the major sources of farmers' information to climate change for majority of the respondents. A little of the respondents claimed that, their awareness of climate change was by their personal experiences. Surprisingly, extension agents who are saddled with the responsibilities of educating farmers in their palaces of assignment particularly in the study area shows little concern on informing farmers on the incidence of climate change.

5. Recommendations

Following the findings and conclusions made from the study, the following recommendations were made:

There is a need for the government and Non-governmental organization to support the participation of female in all agricultural activities through establishment of female farmers' cooperative group or society. By so doing, it will go a long way in improving their standard of living.

Educated persons should be encourage to participate more on agriculture through giving loan and other incentives to under the activities. This will help in improve their standard instead to wait for white collar job which has no time to come.

More emphasis on dissemination of climate change issues through the mass media is highly needed through public-private partnership on adaptation strategies to reduce factors that aggravate climate change and as well as showcase successful adaptation techniques adopted elsewhere that have local relevance.

Also there is need for the government to motivate extension agents to continue discharging their public responsibilities in order to improved farmer education for effective agricultural activities and adaptation strategies on the changing climate.

There is need for public-private partnership to provide financial support to farmers in order to increase their ability to adopt various management strategies in response to climate change.

Author details

Salisu Lawal Halliru[1*], Abubakar Abdullahi Bichi[2] and Aliyu Shu'aibu Muhammad[1]

1 Department of Geography, Federal College of Education, Kano, Nigeria

2 Department of Geography, Federal University, Gusau, Nigeria

*Address all correspondence to: lhsalisu09@gmail.com

References

[1] Synthesis Report. Contribution of Working Groups I, II and III to the Fourth Assessment Report of the Intergovernmental Panel on Climate Change [IPCC]: Climate Change 2007 Geneva, Switzerland: IPCC

[2] Mcquire B, Mason I, Kilburn C. Natural Hazards and Environmental Change: London: Arnold, 53-63 p

[3] Olaniran, O. J. Rainfall Anomalies in Nigeria: The Contemporary Understanding: The Fifty-Fifth Inaugural Lecture of the University of Ilorin.

[4] Odjugo P.A.O & Ikhuoria A. I. The impact of climate change and anthropogenic factors on desertification in the semi-arid region of Nigeria: Global Journal of Environmental Science, 2[2]: 118-126.

[5] Ayoade, J.O. Climate change: A Synopsis of Its Nature, Causes, Effects and Management: Ibadan: Vantage Publishers, PP. 5-6.

[6] Odjugo P.A.O. An analysis of rainfall pattern in Nigeria: Global Journal of Environmental Science, 4(2): 139-145 p

[7] Nwafor, J. C. Environmental Impact Assessment for Sustainable Development: The Nigerian perspective. Enugu: EL 'DEMAK pubs, Pp. 359-394 p

[8] Mshelia A.D. Adaptation strategies to climate change: Journal of Energy and Environment, 18 (3): 74-81 p.

[9] Climate change impact on plant species composition in six semi-arid rangelands of Northern Nigeria: Nigerian Geographical Journal 5[1]: 35-42 p

[10] Anyoha, N. O., Nnadi, F. N., Chikaire, J., Echetama, J. A., Utazi, C. O., & Ihenacho, R. A. Socio-economic factors influencing climate change adaptation among crop farmers in Umuahia South Area of Abia state, Nigeria: Net Journal of Agricultural Science, 1[2], 42-47 p

[11] Nwajiuba C, editor. Climate change and adaptation in Nigeria: Margraf, Weikersheim, 65-72 p

[12] Sofoluwe, N. A., Tijani, A. A. and Baruwa, O.I. Farmers Perception and Adaptation to Climate Change in Osun State Nigeria: African Journal of Agricultural Research. Vol. 6 [20]. Pp. 4789-4794 p

[13] Tanko and Momale, editors. Weather and Climate: Kano environment, society and development. 2014. 13-19 p. United Kingdom: Adonis & Abbey Publishers Ltd

[14] Census: National Population Commission

[15] E. A. Olofin, Nabegu and Dambazau, editors. The Physical Setting: Wudil within Kano Region. A Geographical Synthesis. 2008. 5-34 p. Kano: Adamu Joji.

[16] Krejcie, R. V., & Morgan, D. W. Determining sample size for research activities in educational psychological measurement: Journal of Education and Psychology, 1970 30, 607-610 p

[17] Abiola, O.Procedures in Educational Research, Kaduna: Kingon Nigeria Limited.

[18] Falola, A and Achem, B. A. Perceptions on climate change and adaptation strategies among sweet potato farming households in Kwara State, North central Nigeria: Ceylon Journal of Science 46 (3): 55-63 p

[19] Adekunle, O.A., Adefalu, L.L., Oladipo, F.O., Adisa, R.S., and Fatoye,

A.D. Constraints to youth's involvement in agricultural production in Kwara State, Nigeria: Journal of Agricultural Extension, 13[1]:102-108 p

[20] Falola, A., Ayinde, O.E., and Ojehomon, V.E. T. Economic analysis of rice production among the youths in Kwara State, Nigeria: Albanian Journal of Agricultural Sciences, 12 [3]:503-510 p

[21] Ikpe, E.Unpublished Master's thesis Adaptation strategies to climate change among grain farmers in Goronyo Local Government Area of Sokoto state. Department of Geography, Ahmadu Bello University, Zaria.

[22] Ishaya, S., & Abaje, I. B. Indigenous people's perception on climate change and adaptation strategies in Jema'a Local Government Area of Kaduna State, Nigeria: Journal of Geography and Regional Planning, 1[8], 138-143 p

[23] Abraham, F., Bamidele, F. S., Adebola, A. J., & Kobe, I. H. Climate change mitigating activities and determinants in the rural guinea savannah of Nigeria: Sustainable Agriculture Research, 1 [2], 170-177 p

[24] Odewumi, S.G., Awoyemi, O. K., Iwara, A.I. and Ogundele, F.O. Farmer's perception on the effect of climate change and variation on urban agriculture in Ibadan Metropolis, South-western Nigeria: Journal of Geography and Regional Planning. Vol. 6 [6], 209-217 p

[25] Muhammad-Lawal, A., Omotesho, O.A., and Falola, A. Technical efficiency of youth participation in agriculture: A case study of the Youth-in-Agriculture Programme in Ondo State, Southwestern Nigeria: Nigerian Journal of Agriculture, Food and Environment, 5[1]:20-26 p

[26] Francis, N. & Tsunemi W. Determinants of farmers' adaptation to climate change: A micro level analysis in Ghana: Scientia Agricola. Vol 73, N.3, 201-208 p

[27] Zhao, X., Leiserowitz, A.A., Maibach, E.W. and Roser-Renouf, C. Attention to science/ environment news positively predicts and attention to political news negatively predicts global warming risk perceptions and policy support: Journal of Communication, 61(4):713-731 p

Chapter 12

Spatio-Temporal Dynamics of Soil Microbial Communities in a Pasture: A Case Study of *Bromus inermis* Pasture in Eastern Nebraska

Taity Changa, Jane Asiyo Okalebo and Shaokun Wang

Abstract

Today's intensified agricultural production is characterized by crop and pasture monocultures, which have a significant impact on soil microbial diversity and abundance. This chapter provides a case study in which the relative importance of brome grass (*Bromus inermis*) monoculture pasture versus intra-site microhabitat diversity is explored using fatty acid methyl ester (FAMEs) assay to delineate the presence and abundance of several classes of soil microbes instrumental in soil nutrient cycling, plant health, plant organic matter decomposition, and soil stabilization. The chapter explores spatio-temporal variability of bacteria, actinomycetes, saprophytes, mycorrhizae, and micro-eukaryotes over two durations (summer and fall) collected using two distinct sampling methods. One of the methods is commonly employed, namely, transect-based, while the other is informed by soil electroconductivity measurements conducted over the entire pasture site from a previous survey.

Keywords: grassland, plant-soil feedback, soil electroconductivity, soil health, soil quality, *Bromus*

1. Introduction

The role of below-ground soil processes as well as plant-soil feedbacks, particularly those facilitated by the interaction between plants and soil microbes, are poorly understood despite having a key role in potentially regulating the productivity of ecosystems [1–4]. Soil microbial communities are critical for soil functioning and health as drivers for biogeochemical cycles [5, 6]. They have been shown to be shaped by the abiotic and biotic environments (e.g., plant and animal communities, edaphic characteristics, water, and climate), management practices (e.g., crop rotations and grazing) and are known to alter the amount, quality, and distribution of organic carbon (OC) within the soil profile—key for soil health [1, 7–10].

Today's intensified agricultural production is characterized by crop and pasture monocultures, which have a significant impact on the soil microbial diversity and abundance. As an example, pastures in Eastern Nebraska grasslands have been shaped by Brome grass (*Bromus inermis*), a perennial grass species that was first introduced from Eurasia into North America in the 1880s as a forage grass and for soil stabilization [11]. The use of this species was promoted after the drought of the 1930s to control soil erosion and stabilize soil banks and ditches in Nebraska and elsewhere [11, 12]. The species has been successful at colonizing and spreading into pastures due its vigorous tillering and dense establishment of rhizomes and rootstocks [11, 13, 14]. In Eastern Nebraska, the species thrives well in clay loams forming even stands and monocultures in pastures [11, 13–15]. Deep roots (up to 2.7 m), debris, and exudates have been shown to influence both soil structure and texture positively through increased organic matter availability as well as binding properties which increase soil water holding capacity [11]. While there have been several studies that addressed the role of brome grass in soil processes and organic matter mineralization such as [13–15], some research has been conducting on its role in shaping the microbial community in pasture sites such as those conducted by Grigera et al.; Lauber et al.; Pereira e Silva et al.; and Segal et al. [16–19].

Understanding factors driving the spatio-temporal variability of soil microbial communities in monocultures can inform the selection of best management practices that can maintain and regulate soil biodiversity and enhance the microbial-soil ecosystem functions in these systems. However, quantifying microbial abundance and diversity have been a challenge over a range of spatial and temporal scales [20, 21]. This is in part attributed to the high heterogeneity in microbial populations found in two adjacent locations that are as close as 10 cm apart [21]. Sampling strategies have been proposed to capture the spatial heterogeneity in soil microbial biota [10]. One strategy is the randomized sampling method that is common among researchers [15, 22] and the other is the application of transects across the site of study guided by physical and biological variables including slope, water gradients, soil pH, or plant productivity [10, 23–25]. Each strategy has its strengths and shortcomings, especially where soils are spatially variable. It is agreed upon however, that besides vegetation, soil characteristics such as soil moisture, soil structure, organic matter, and temperature are some of the main drivers that govern spatio-temporal variability in microbial populations [21, 26].

Soil microbes can thrive under a wide range of temperatures. Nevertheless, microbial enzymatic catalytic functions perform optimally over a shorter range of temperatures. During spring and summer seasons (approximately 20–30°C), microbial abundance and activity are both at high levels compared to that measured during winter conditions (>10°C) characterized by reduced vegetation and less favorable edaphic conditions [17, 18]. Low temperatures affect the plant development resulting in the reduction of substrate (nutrients within the rhizosphere of plant roots) for soil microbes [17, 18]. Winter conditions characterized by both low temperatures and low nutrient availability trigger bacterial spore and fungal sclerotia formation. Soil microbial dormancy reduces the microbial abundance and ecosystem functions. Spores, sclerotia, and other soil microbial growths can remain dormant for long durations until the edaphic conditions become more favorable for reproduction and growth. During these periods of low temperatures, enzymatic activities and microbial functions are minimal [17, 18]. In order to understand temporal variation in soil microbial communities and their ecosystem functions, sampling over different temperatures and/or seasons is recommended and advisable.

Physical and chemical properties of soils in monocultures can potentially serve as a guide in delimiting zones of similar soil microbial abundance and diversity compared to sampling along a transect line. A property widely used to monitor soil characteristics is the apparent electrical conductivity (ECa), which measures the amount of electric current that soil can conduct [27]. This valuable trait relates to soil texture, that is, soils with high clay content exhibit higher ECa values and vice versa [28]. The percentage of clay particles in soil influence its physical texture. Additional soil properties such as salinity, water, organic matter content, and cation exchange capacity (CEC) are also directly correlated with ECa. This soil characteristic is used for delineating homogeneous zones and is utilized by large commercial producers for guiding the fertilizer application and other land management practices [27].

This case study assesses the relative importance of brome grass monoculture versus intra-site microhabitat diversity in determining the microbial abundance and diversity. We hypothesize that soil microbial communities are in most part shaped by the brome monoculture rather than the intra-site variability in edaphic factors such as temperature, pH, N, ECa, and OC in a relatively homogeneous site. We tested two soil sampling techniques, ECa-guided sampling technique and transect-based sampling methodology, and used fatty acid methyl ester (FAMEs) assay [19, 29, 30] to determine and quantify the soil microbial diversity and abundance. The assay characterizes the composition of fatty acid species recovered from membranes of soil microbes which serve as unique markers for soil microbiota identification [30]. FAMEs assay is cost-effective and has been shown to give similar results to genomic analysis of bacterial and archaeal ammonia oxidizers [19].

2. Materials and methods

2.1 Site description

The experiment was conducted at a 15-ha pasture site, located at the Eastern Nebraska Research and Education Center (ENREC), University of Nebraska-Lincoln (latitude 41° 8′28.45″ and longitude N, 96°27′1.05”W) (**Figure 1**).

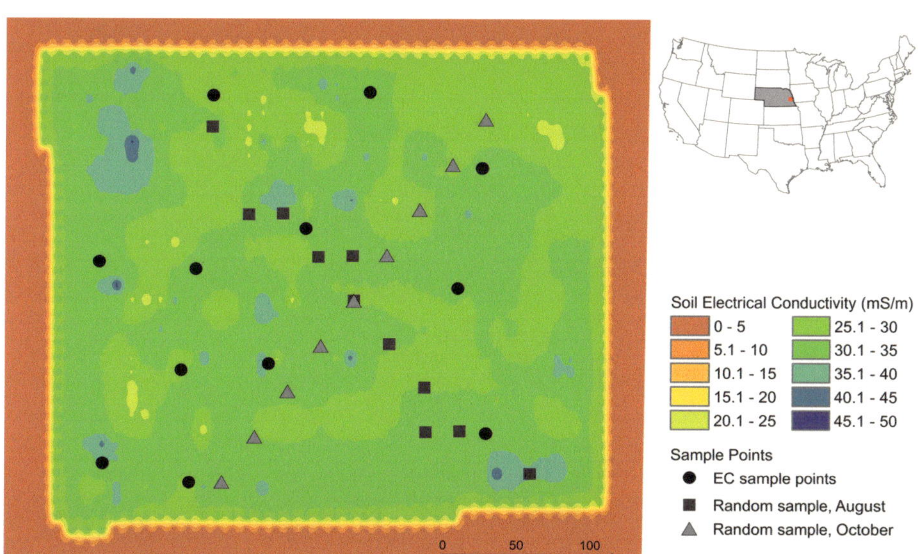

Figure 1.
Map showing distribution of soil apparent electrical conductivity (EC_a) mS m^{-1} in brome grass pasture study site. The points on the map indicate locations where soil sample were taken following EC (●) and random August (■) and October (▲)-based techniques. On the right corner is the site location in highlighted state of Nebraska and continental map of USA.

Annual average precipitation is 748 mm and the mean annual temperature is 9.9°C. January and July are the coldest and warmest months of the year with an average minimum of -11.3°C and maximum of 30.4°C, respectively (https://www.usclimatedata.com/climate/mead/nebraska/united-states/usne0316). The precipitation in 2017—when sampling took place—totaled 869 mm, with 112.3 mm in the month of August and approximately 24.4 mm falling 6 days prior to soil sampling on August 31. Minimal rainfall amounts of less than 2.54 mm had been recorded 11 days prior to sampling on 27th of October. The total amount of rainfall in October was 128.3 mm with most of it falling in the first half of the month.

The site is ~360 m above sea level with a topography that is relatively flat (0–2% slope). Under the older soil classification, the soil is characterized as sharpsburg silty clay loams of the fine, smectitic, mesic Typic Argiudoll series. Under the new classification, the pasture is comprised of yutan silty clay loam, tomek silt loam, and filbert silt loam with each occupying about one third of the pasture.

The soils are slightly acidic (pH 5.8) and are characterized by a high organic matter (OM 4.1%) (**Table 1**). The pasture was seeded over 20 years ago with smooth brome grass *Bromus inermis* Leyss which forms a monoculture. Brome grass is

Soil chemical properties	Value
1:1 Soil pH	5.80
WDRF Buffer pH	6.80
1:1 S Salts mmhocm^{-1}	0.09
Organic matter LOI %	4.14
Nitrate-N ppm N	4.42
Potassium ppm K	462.40
Sulfate-S ppm S	34.40
Zinc ppm Zn	2.81
Iron ppm Fe	189.38
Manganese ppm Mn	31.68
Copper ppm Cu	1.95
Calcium ppm Ca	2452.00
Magnesium ppm Mg	579.80
Sodium ppm Na	35.80
Boron ppm B	0.55
CEC/Sum of cations me/100 g	20.12
%H Sat	8.80
%K Sat	6.00
%Ca Sat	60.80
%Mg Sat	23.60
%Na Sat	0.80
Chloride ppm Cl	8.52
Mehlich P-III ppm P	30.40

Table 1.
Physical and chemical properties in the brome grass pasture site in eastern NE. (Ward Laboratories, Kearney, Nebraska, 2017).

characterized with low C:N ratio of about 60 [31]. In studies conducted by Vinton and Goergen [32], brome grass yielded a C:N ratio of 50.68 compared a high value (102.1) for *Panicum virgatus* (switchgrass) which makes it highly palatable for livestock [32]. It grows rapidly and aggressively at the start of the growing season and does well in soils that are rich in nitrogen. Its low C:N ratio results in relatively fast decomposition rates thereby increasing soil available N and rapid expansion and growth of the species in a positive plant-soil feedback system. Every year in spring, nitrogen fertilizer is applied at the rate of 14.6 kg ha^{-1}, and since 2014, animals (cattle) graze on this pasture between the end of April to end of September. N fertilization has been conducted as a management practice in the north central USA to improve beef animal daily gain (ADG) through increased forage biomass in pastures attributed to N availability through enhanced N cycling processes [33].

2.2 Apparent electrical conductivity (ECa)

A soil survey to characterize the soil ECa at the pasture site was conducted on April 4, 2017. The survey consisted of 22 transects running from north to south with an average spacing of 15 m between transects. A Geonics dual-pole EM38 (Geonics, Inc., Mississauga, Ontario, Canada) was mounted in a plastic sled and towed through the fields at about 5 km $h.^{-1}$ using a four-wheel all-terrain vehicle. Soil ECa sampling locations were determined using a Geode global positioning system (GPS) (Juniper Systems, Sunnyvale, CA, USA). The geographical coordinates as well as EM38 output were collected every 10 s and stored on a data logger. The raw data was processed using the ESAP-95 software [34] to identify 12 sampling points that best represented the spatial soil ECa variability within the field.

2.3 Soil physical and chemical properties

A set of samples used for the determination of the soil physical and chemical properties were collected in early June. These samples were collected at a depth of 20 cm and included soil cores taken from five randomly selected locations in the pasture to represent field scale soil properties. Samples were air-dried, sieved to remove large debris, and bulked into five composite samples and sent to Ward Laboratories Inc. (Kearney, NE.) for soil analysis (**Table 1**). The second set involved soil cores obtained from the designated 12 points that were chosen to represent the soil ECa gradient of the pasture as described above in the pasture ECa survey. Soil cores were collected from 10 to 15 cm depth increments and analyzed as follows: substrate water content was determined gravimetrically, samples were weighed, oven-dried at 105°C for 2 days, and weighed again. Laboratory EClab and pH were determined in 1:1 water:substrate slurries using a conductivity meter for EClab and a glass electrode for pH [35]. Extractable phosphorus (P) was determined using the method of [36] with P concentration determined spectrophotometrically at 882 nm using the phosphomolybdate blue method [37]. Inorganic nitrogen (N) in 1 M KCl extracts was measured colorimetrically using an AA3 flow injection ion analyzer (Seal Analytical, Inc. Mequon, WI). Nitrate-N was determined using the Cd reduction method [38]. Total C and N were measured by dry combustion (EA1112 Flash NC Elemental analyzer, Thermo Finnegan Scientific Inc., Waltham, MA) of air-dried ground soil.

2.4 Fatty acid methyl ester (FAMEs) analysis

Soil samples used for quantification of soil microbial community and abundance were collected in August at peak biomass and October when regrowth occurred. Measurement of ECa was conducted at the beginning of the season prior to livestock being grazed on the pasture field. ECa is a product of dynamic soil factors (e.g., soil moisture) and static measurements (e.g., bulk density, clay type) [39] and is generally stable throughout the growing season. Sampling following the soil ECa gradient (described above) was repeated at the 12 selected points, while soil sampling following the transect method were taken at points about 15–25 m apart along a SE-NW and NE–SW transects during August and October sampling, respectively. The GPS coordinates of locations where soil samples were collected were obtained via the GPS app locator of a smartphone to identify transect sampling points on the pasture site map. Approximately, 50 g of soil core taken at a depth of 10–15 cm was transferred to plastic ziplock bag and kept in an icebox. Between samples, the soil corer was cleaned with alcohol (70%) to prevent cross contamination of samples. Soil samples were transported to the laboratory where they were processed using 2 mm sieves to remove pebbles and any plant material and immediately stored in the freezer (-20°C) until ready for assay.

For the assessment of microbial diversity and abundance, the total microbial fatty acids (FA) were extracted following the procedure described by Schutter and Dick [30]. Briefly, total microbial lipids in 5 g soil samples were extracted in 10 ml of 0.2 M methanolic potassium hydroxide and the mixture heated at 37°C for 1 h with intermittent shaking. The solution was then neutralized by adding 1 N acetic acid and the lipids dissolved in hexane. The mixture was centrifuged at 6000 rpm for 10 min and the supernatant was carefully recovered, filtered, and further processed before fatty acids were quantified using gas chromatography with 0.05 mg/ml nonadecanoic acid (C19:0) as an internal standard. A total of 19 FAMEs were retained and used to determine microbial community composition following FAMEs nomenclature of the IUPAC-IUB Commission on Biochemical Nomenclature (IUPAC-UIB, 1987). Specific fatty acids with 14–20 carbon composition were used to represent fungal, bacterial groups, and micro-eukaryotes. Bacterial biomass was represented by the sum of 10 FAMEs: iC14:0, iC15:0, aC15:0, C15:0, iC16:0, iC17:0, aC17:0, C17:0, cyC17:9, cyC19:9,10, and cyC19:11,12 [16]. Actinomycetes bacteria were quantified by 10Me fatty acids: 10MeC18:0 and 10MeC19 [22, 40], while saprophytic fungal biomass was represented by C18:2cis9,12 [41]. In addition, micro-eukaryotic biomass was represented by the sum of C20:3, C20:4 and C20:5 [19, 42]. Finally, the fatty acid C16:1cis11 was used as a biomarker for arbuscular mycorrhizal fungi (AMF) [43]. Total microbial biomass was estimated by summing up FAMEs representing bacteria, actinomycetes, saprophytic fungi, and AMF. In addition, total FAMEs for bacterial (bacteria and actinomycetes) and fungi (AMF and saprophytic fungi) were used to calculate the ratio of fungi to bacteria biomass in the soil.

3. Statistical analysis

The resulting datasets from FAMEs analysis on the microbial groups were analyzed using generalized linear mixed models procedure (GLIMMIX Procedure) in SAS® 9.4 software package and means separated at $p < 0.05$. This analysis involved testing sampling methods (ECa-directed vs. transect-based), the effect of sampling date/time as well as any interactions between sampling method and time. To further investigate any impacts of plant (brome grass monoculture) as well as the effect of soil physical and chemical parameters on microbial diversity and abundance, principal component analysis (PCA) was conducted. The PCA was performed to determine the contribution of soil characteristics (e.g., ECa, pH) in the variation and separation of both temporal and sampling method. Soil chemical, physical and biological attributes were used in the PCA to elucidate the effects of sampling technique and timing of sampling. This was conducted using R programming software version 3.4.1, utilizing ade4 package [44] and factoextra package for visualization purposes [45].

In addition, the relationship between soil microbes and soil physicochemical characteristics was visualized using heatmap. The heatmap was generated using a combination of packages and their functions in the R programming language. The hclust function and scale both available in stats package were utilized, respectively, for hierarchical cluster analysis after applying the scale function to centralize the various data about the mean and to generate z-scores around each variable's standard deviation about the mean. Cluster analysis was performed using the default 'complete' agglomeration method. The melt function in the reshape2 package was used to organize the dataset before plotting the heatmap using ggplot function acquired from the ggplot2 package. The z-score legend was generated using scale_fill_gradient2 function in ggplot2 package and color breaks represented z-scores ranging from -3 to 3 over the entire dataset.

4. Results and discussion

4.1 Site characteristics

The physical and chemical characteristics of soil in the brome grass pasture are summarized in **Table 1**. Soil organic matter contents were relatively high, averaging 4.14%. High organic matter content in the top 20 cm of the pasture soil can be attributed to dense rhizomatous roots of brome grass root biomass and shoots [32]. Soil nutrient availability to crops is influenced by soil pH. The pasture site exhibited a slightly acidic soil pH (5.98) and was positively correlated (R^2 0.89, $p < 0.05$) with calcium but negatively correlated to percent H+ (R^2 0.95, $p < 0.05$). Available N (Nitrates-N ppm) of soils sampled from the pasture site averaged 4.42, which can be attributed to the cow dung manure, N fertilizer, as well as the high biomass from the brome grass shoots and roots. 1:1 soil soluble salts (mmho cm^{-1}) was strongly and positively correlated to measured available soil nitrates (R^2 0.79, $p < 0.05$). Available P in the pasture site was measured at 18.07 ppm P. Hydrogen (H^+) cations contributed significantly to the total sum of cations me 100 g^{-1} (CEC) of the pasture site.

Values for ECa ranged from 21 to about 44 mS m^{-1} (**Figure 1**), which indicated a low to moderate level of spatial site heterogeneity. The mean ECa value was 32 mS m^{-1} with ECa of 30.1–35 and 25.1–30 mS m^{-1} being more common (the two combined covered nearly the entire the pasture). Regions with lowest and highest ECa values mainly constituted small pockets that were randomly distributed across the pasture with no clear pattern that could be discerned (**Figure 1**). As a result, soil in this pasture was considered to be less variable and the ECa values fell within normal range of 0–150 mS m^{-1} for grass pasture [35] but lower than in a fertilized maize field [46] within Eastern Nebraska.

4.2 Soil microbial community

General composition of microbial communities, namely, total microbial biomass, diversity, and composition of soil microbes detected using FAMEs assay in soils sample collected over two seasons and methods is presented in **Figures 2** and **3**. Each sector of the pie chart (**Figure 3**) represents individual microbial composition as a percentage of total recovered microbial fatty acid. Microbial biomass was dominated by bacteria (55–65%), followed in declining order by arbuscular mycorrhizae (15–25%) saprophytic fungi (8–9%) actinomycetes (8%) micro-eukaryotes (4%).

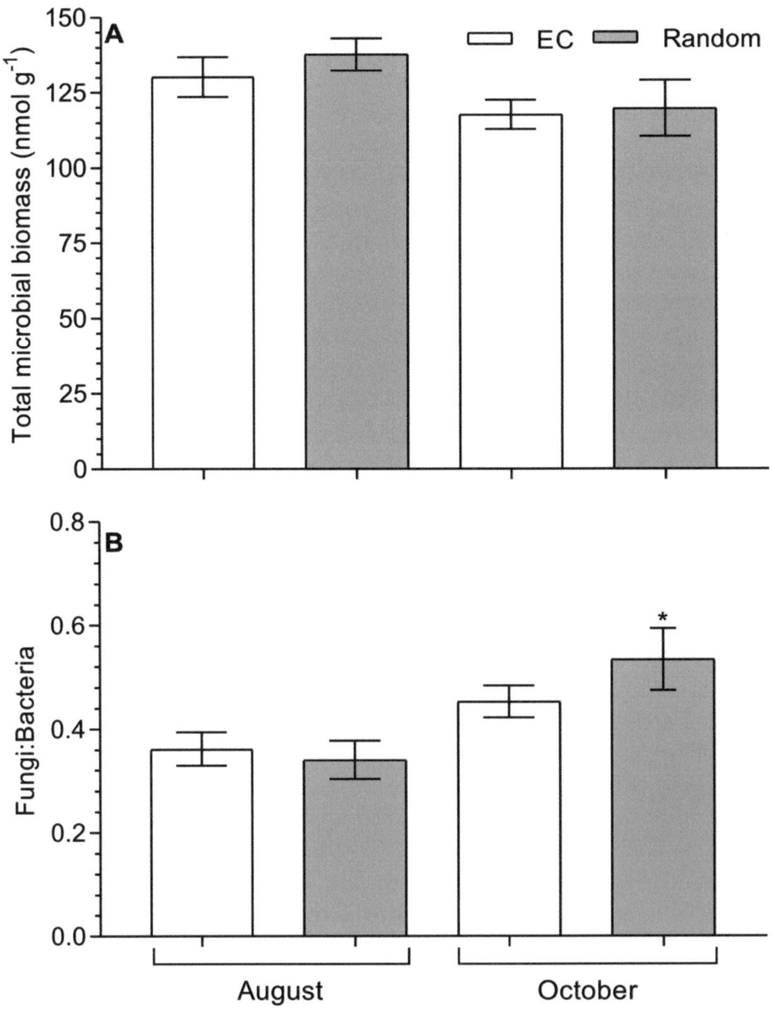

Figure 2.
Total microbial biomass (A) and the ratio of fungi to bacteria (B). Individual bars represent the mean and standard error collected following EC_a (clear bars) and random method (dark bars).

Bacteria was found to be highly correlated (R^2 0.84, $p < 0.05$) with actinomycetes. This was consistent across the two sampling durations. Bacteria correlation with saprophytic fungi (R^2 0.53–0.6, $p < 0.05$) and micro-eukaryotes (R^2 0.56–0.59, $p < 0.05$) was only significant during August and October sampling, respectively. The abundance of soil micro-eukaryotes showed significant correlation with AMF (R^2 0.74–0.77, $p < 0.05$) for both soil sampling methods while the correlation with saprophytic fungi (R^2 0.64, $p < 0.05$) and actinomycetes (R^2 0.72, $p < 0.05$) was observed only in August and October, respectively. Furthermore, the total recovered microbial FA was found to be strongly correlated with all microbial groups ranging between R^2 0.59 and 0.89 and was statistically significant across all microbial groups and sampling times except for AMF ($p = 0.59$) in October samples. The high correlation of bacteria with actinomycetes are similar to that of Grigera et al. [16] who established a high level of correlation (R^2 0.88, $p < 0.05$) in an agricultural field in Buffalo county, Nebraska, continuously cropped with corn. These high correlations highlight similarities in the edaphic conditions (e.g., pH and organic matter content) in which the microbes coexist in complementary biogeochemical functions in recycling both N and C [47].

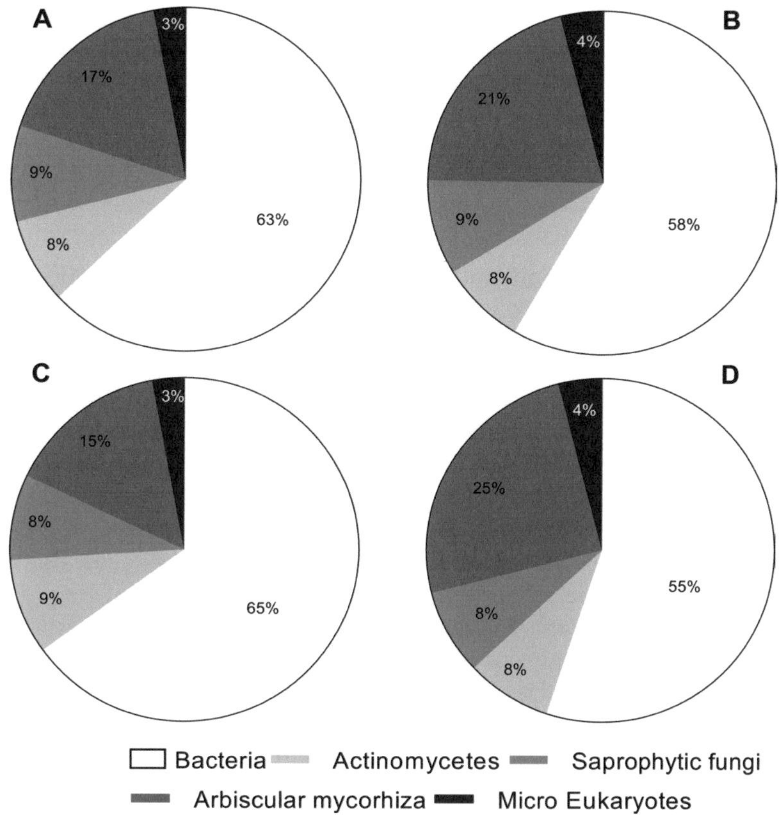

Figure 3.
Pie charts showing diversity and composition of soil microbes detected using FAMEs assay in soils sample collected in august (a and C) and October (B and D). Soil samples were collected following ECa-based method (A and B) and random method (C and D). Each sector represents individual microbial composition as a percentage of total recovered microbial fatty acid.

4.3 Comparison of sampling methods

Statistical analysis did not show any significant differences in soil microbial biomass of soil samples collected based on ECa stratification or transect sampling methods. The sampling method did not result in any significant differences (**Table 2.**) in the abundance of bacteria, actinomycetes, saprophytes, mycorrhizae, or micro-eukaryotes. Transect sampling technique has the same sensitivity and reliability as an ECa-based method in capturing the spatial and temporal dynamics of soil microbiota and can thus be used as a method of choice for sites with a relatively low range of ECa variability indicative of similar soil chemical,

	Type III tests of fixed effects				
Effect	**Bacteria**	**Actinomycetes**	**Saprophytes**	**Mycorrhizae**	**Micro-eukaryotes**
Date (D)	<.0001	0.002	0.239	0.076	0.740
Sampling (S)	0.583	0.650	0.660	0.413	0.589
DxS	0.257	0.274	0.656	0.318	0.951

Table 2.
Summary table of date (D) and sampling methodology (S) and their interactive effects of D and S on bacteria, actinomycetes, saprophytes, mycorrhizae, and micro-eukaryotes sampled at the PR-HPA site in Mead, Nebraska.

physical and microbial properties. The two soil sampling techniques used in this research (i.e., ECa- and transect-based) captured comparable soil microbial communities and abundance in both space and time highlighting the significant role of vegetation on soil microbial communities as highlighted by others like Grigera et al.; Lauber et al.; Pereira e Silva et al.; and Segal et al. [16–19].

4.4 Effects of temperature on soil microbes

There were statistically significant seasonal differences in the total soil microbial biomass irrespective of soil sampling technique with a considerably higher abundance in August compared to October. Total soil microbial mass in August and October soil samples had means of 130.2 and 137.7 nmol g^{-1} in August compared to 117.7 and 119.8 nmol g^{-1} in October for soil samples collected via ECa-directed and transect-based sampling methods. Despite the observed decline in October, which was cooler, no statistical difference was observed in space and time for microbial biomass.

When examining the effect of sampling date (i.e., August vs. October) on the abundance of individual microbial groups, results demonstrated a significant shift in soil biota with temporal changes affecting selective groups. Specifically, saprophytic fungi, actinomycetes, and micro-eukaryotes remained seasonally stable and constituted about 8–9 and 3–4% of the total microbial biomass, respectively (**Figure 2**). Bacteria and AMF abundance exhibited a significant temporal variability. In particular, a significant decline of bacterial biomass ranging from 4 to 10% observed in August and October, respectively, was observed irrespective of method of soil sampling. In contrast to bacteria, a significant increase (4–10%) in AMF abundance was noted in soil sampled in October compared to August (**Figure 4**). Bacterial abundance declined by up to 10% from August to October with a corresponding 10% increase in AMF observed during the same period.

Fungi to bacterial (F:B) ratio which is indicative of the changes in soil microbial communities [48] was calculated in August (0.30–0.39) and noted to have increased (0.45–0.53) in October, representing a 20 and 32% increase for ECa and transect soil samples, respectively. Although there was a general increase in F:B in October samples, statistical significance was only observed in soil sampled via the transect

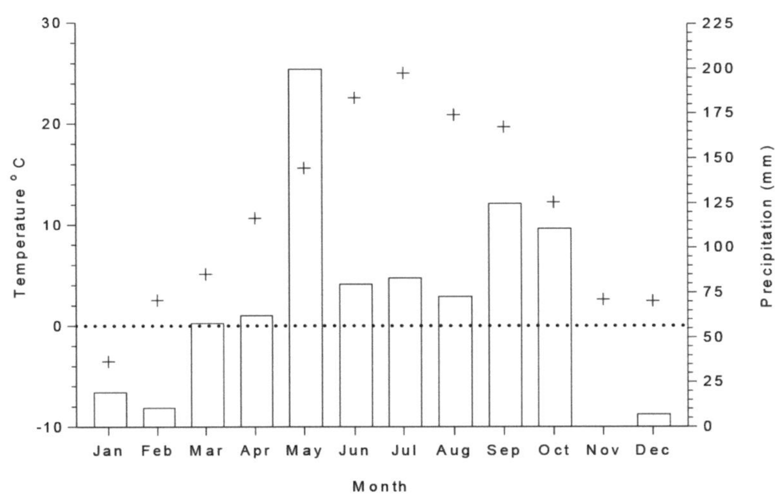

Figure 4.
Monthly average temperature and precipitation from Mead weather station (41.17° N, 96.47° W) closest to the pasture study site (4 km) at the East Nebraska research and Experimental Station (ENREC). Bars indicate total monthly precipitation while stars show mean monthly temperature.

method (**Figure 3B**). Increase in F:B ratio during the cooler month reflected the shift in compositional abundance of fungi and bacteria that was observed in October (**Figure 2**). Commonality in trend in total microbial biomass and the ratio of fungi to bacteria as observed along time illustrates the comparable sensitivity of the two soil sampling methods [16, 22, 49].

Shifts in soil microbial communities are affected by seasonality and specifically temperature and moisture. Temporal changes in soil microbial abundance observed in our work has also been demonstrated elsewhere in soil under controlled environment [3, 50] as well as in different ecosystems including forests [51–54], deserts [55], cultivated land, [17–19] as well as grasslands [17]. While fluctuation in microbial abundance across the soil types are common, and the key drivers tend to vary according to the location and soil type, with environmental factors such as temperature and precipitation being the most dominant factors [17, 50, 54, 55].

Our results concur with those of several researchers such as Papatheodorou, Argyropoulou, and Stamou [56] who conducted studies on soils from a grassland in Mediterranean Greece. They detected a linear decline in bacterial diversity, evenness, richness, and mean oxidation, especially of carbohydrate and carboxylic substrates over a 6-month (July to December) study. Substrate use efficiency and specifically carbon use efficiency have also been found to decrease with nutrient availability and increasing temperature [57]. Our study's August and October environmental conditions showed that the monthly mean temperature was higher (20.8°C) in August compared to October (12.2°C). Cumulative monthly precipitation measured 78.89 mm and 110.99 mm for August and October, respectively, which for this area is about average for August but double the average for the month of October (**Figure 2**). In addition, we examined in detail these two environmental parameters recorded 4 days preceding the soil sampling dates. The results summarized in **Table 3** showed that average temperature was 20.5°C in August and dropped to 10.7oC in October. With respect to precipitation, a cumulative 1.27 mm of rain was received 3 days prior to the August sampling and none prior to the October sampling. While there were differences in precipitation both monthly (August and October) and the days preceding the aforementioned sampling dates, this variation did not have a significant effect on soil moisture as revealed by computed soil gravimetric water content (**Table 3**). This implies that, changes observed in bacterial and fungal composition (**Figure 4**) may possibly have resulted from factor(s) other than soil moisture which are discussed in the next section below. These results concur with those of [58] who characterized soil microbial communities and their conditioning by varied plant species. They noted that soil bacterial communities are primarily influenced by abiotic conditions; namely temperature and ECa (**Figure 5**). Fungal communities on the other hand are determined by biotic conditions such, as plant species [58] as seen with the

Date	Sample type	Temperature (°C)	Precipitation (mm)	Gravimetric water content (%)
August 31	EC	20.5 ± 2.2	0.3 ± 0.4	18.7 ± 5.9
	Random			19.8 ± 6.2
October 27	EC	10.7 ± 2.5	0	17.84 ± 2.4
	Random			17.39 ± 2.1

Table 3.
Summary of weather data and soil water content. Values indicate means and standard deviation of temperature and precipitation recorded 5 days before soil samples were collected. Gravimetric soil water content was calculated as the difference between fresh and dry soil of a unit of soil and the values indicate mean and standard deviation.

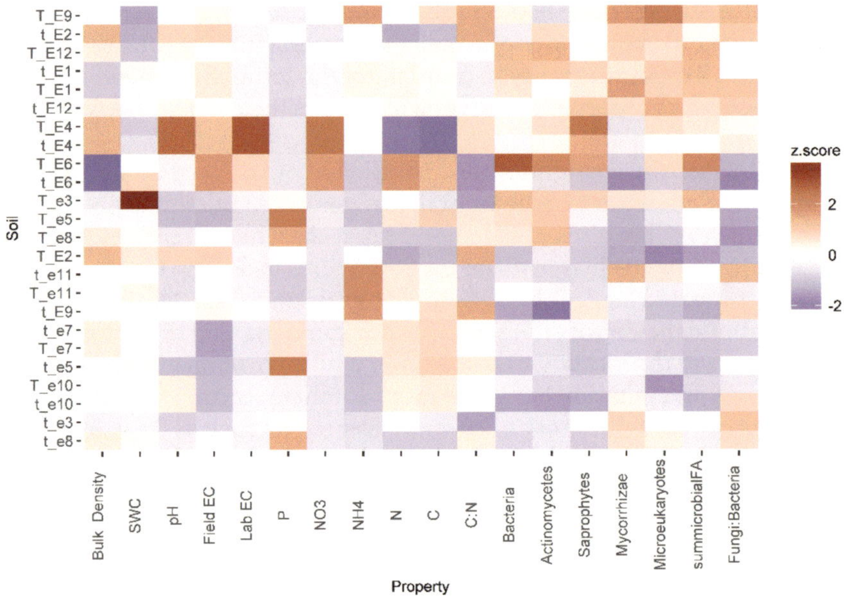

Figure 5.
Cluster heat map showing relationship between soil microbiota, soil physicochemical attributes, and environmental variables. Soil properties and microbes are indicated on the bottom, while soil EC_a and temperature are to the left of the image. The rows representing soil samples characteristics are clustered based on hierarchical cluster analysis of the values of the measured soil variables represented in the columns. These soil variables are sorted based on physical, chemical, and biological characteristics. Letters "E" and "e" represent high and low field EC_a, respectively, and classification was based on the median EC_a value of 32.5 mS m^{-1}. The EC_a values greater than 32.5 mS m^{-1} units are represented by "E," while those below the media are denoted by "e." Letters "T" and "t" represent the warmer and cooler month of August and October, respectively. The z values are represented by the blue color, while the color intensity shows the level of significance.

increased flush of Brome grass root growth during the late fall season. The brome plant-AMF microbial feedback elicits subsequent biases toward the development of brome grass monoculture.

4.5 Impacts of chemical, biological, and physical properties on soil microbial communities

The overall relationship between the soil microbes and soil physiochemical characteristics was computed and summarized in https://prhpa.unl.edu/supplementary-materials-ltar-pasture-soil-characteristics-0. The results show significant correlations between soil physical, chemical, and biological characteristics. In general, there were 30 and 24 statistically significant correlations among the measured soil parameters sampled in August and October, respectively. A total of 15 of these correlations were consistent across the two sampling time points (https://prhpa.unl.edu/supplementary-materials-ltar-pasture-soil-characteristics-0).

Concerning microbial groups and their abundance as impacted by soil physicochemical characteristics, bacteria was negatively correlated with bulk density (BD) and NO_3^- at −0.74 and − 0.66, respectively, while being positively correlated (R^2 0.73) with C:N. In addition, saprophytic fungi were strongly correlated with ECa, EClab, and NO_3^- at R^2 0.6, 0.84, and 0.88, respectively. On the other hand, in soil sampled in October, statistical significance was solely observed between AMF and soil NO_3^- (R^2–0.61). These results are in agreement with the direct effect of BD on soil drainage and its negative influence on bacteria populations similarly observed in crop-livestock studies [59]. Additionally, soils with high available NO_3^- demonstrate a lower population of bacteria necessary in nitrification processes

[47]. Soils with low N content, indicative of high C:N ratios, have higher bacterial populations. A highly positive association of ECa, EClab and NO_3^- with saprophytic fungi is attributable to the cationic byproducts as well as NO_3^- from extracellular breakdown of organic matter by the saprohytic fungi's enzymes. Seasonal variations that lead to lower soil temperatures influence bacteria populations and nitrification processes reducing NO_3^-. On the other hand, AMF colonize roots of regrowth brome grass roots resulting in their higher prevalence and abundance in October soil samples. Our findings concur with those of [58] who determined that plant species determined the relative abundance of AMF fungi in comparison to saprotrophs (e.g., saprophytic fungi and bacteria) which were influenced by soil abiotic factors such as pH.

The relationship between soil microbes and soil physicochemical characteristics as impacted by ECa variability of individual sampling point was examined using PCA (**Figure 6**). The first two PCA axes explained a total of 57.4% of the variability with PC1 and PC2 explaining 32.3 and 27.4%, respectively, of the total variation between several of the soil physicochemical characteristics, temperature, and microbial diversity (**Figure 5**). Specifically, PC1and PC2 largely explained variability in microbial groups and soil characteristics, respectively, (**Figure 6**). A consistent discrimination of soils was evident in the sum of microbial fatty acids which was noted as being negatively associated with relatively low ECa levels. High levels of ECfield, EClab, and NO_3^- where associated with soils of high microbial abundance which was similarly observed by [16] in a corn field in Buffalo county in Nebraska. This observation is also noted in this study as shown in the right hand corner of the heatmap which generally depicts soils of 32.5 mS m^{-1} exhibiting z-scores of above 0 across microbial types (**Figure 5**). High P levels on the other hand decreased AMF root colonization and spore density thereby decreasing microbial abundance and diversity in soils that are relatively high in P from sources such as fertilizers [60, 61]. Sources of P in this pasture site included effluents from the livestock in the form of manure and urine.

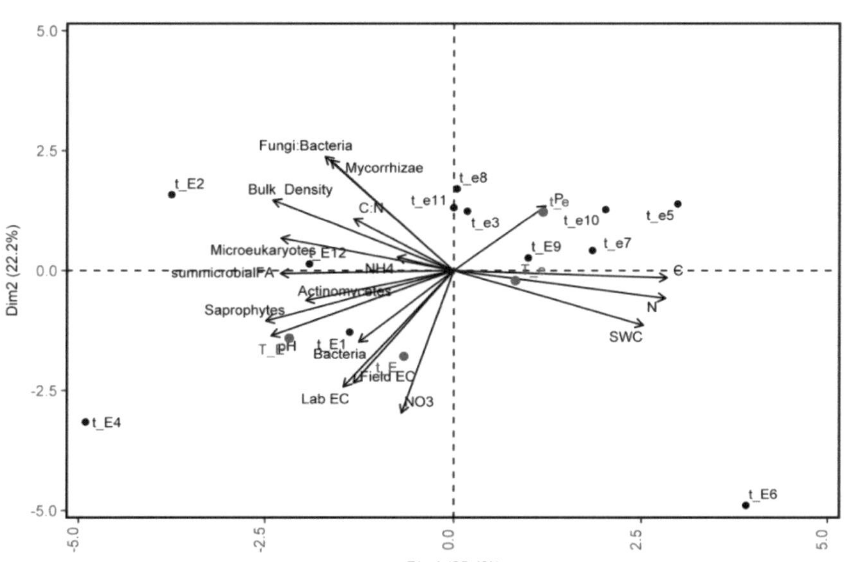

Figure 6.
Principal component analysis (PCA) of microbial groups in soil sampled via ECa method during the warm (August) and cooler (October) temperature. Abbreviations containing a combination of letters and numerals denote the seasons temperature (warm "T" and cool temperature "t") followed by soil ECa (high "E" or low "e" soil ECa values relative to the median value of 32.5 mS m^{-1}), while the numeral (1–12) indicates the point on the field where soil samples were collected following ECa gradient.

The pasture site comprised of a monoculture of brome grass, a cool season grass species with extensive below ground rhizomes and a unique a capacity to maintain active growth during cooler weather. Brome grass has been reported to yield high root biomass [62] with approximately 1014 g m^{-2} root mass measured in 0–10 cm depth of the soil. The seasonal shifts impacts in root biomass production impacted soil microbial communities specifically increasing AMF abundance [63] by influencing C availability [52, 64]. We speculate that as these plants were undergoing late season growth, facilitated by their inherent ability to withstand low soil temperatures (can withstand temperatures as low as −28°C); C allocation to the rhizomes (storage organs) increased as a survival strategy thereby affecting root exudate production. Exudates acting as cues coupled with changes in the production of these compounds have been shown to impact soil microbial community and composition [65]. Thus, elevated production of these signal molecules may have triggered a surge in species of AMF that preferentially associate or benefits from this grass species [66–68]. Bacteria in turn are critical for C nutrient cycling [69, 70]. Some bacteria species are also known to interact with plants symbiotically while fixing nitrogen and also externally in root zones, decomposing organic matter and releasing nutrients to plant roots [71]. Their abundance is largely influenced by the substrate quality of the roots and their exudates.

5. Conclusion

The underlying mechanisms that influence feedbacks and vegetation dynamics within a complex plant-microbial community interaction are largely unresolved [4, 56]. Soil ecosystems are dynamic and diverse, and their physicochemical characteristics vary spatially and temporally. In this study, we compared and contrasted the intra-microbial abundance and diversity of a pasture site in two sampling periods and sampling methods. Our results showed that several classes of soil microbes instrumental in soil nutrient cycling, plant health, plant organic matter decomposition, and soil stabilization were present. These included in order of abundance: bacteria (63%), AMF (17%), saprophytic fungi (9%), actinomycetes (8%), and micro-eukaryotes (3%). The composition of the soil communities changed with the falling temperature, with bacterial abundance diminishing by up to 10% from August to October with a similar magnitude of increase in AMF observed during the same period. Our results showed that the soil bacterial communities were primarily influenced by abiotic conditions, while fungal communities were shaped by the biotic environment such as the plant species such as seen during the flush of regrowth by brome grass (cool season grass) and reallocation of nutrients to root growth that contributed to AMF rapid proliferation and abundance. These findings may provide reasonable evidence that a prolonged positive feedback between brome grass plant-AMF microbial interactions elicits subsequent biases toward the continued dominance and development of brome grass monoculture in a site that was once natural grassland.

Our findings showed that the random sampling technique has the same sensitivity and reliability as an ECa-based method in capturing the spatial and temporal dynamics of soil microbiota and can thus be used as a method of choice for sites with a relatively low range of ECa variability, indicative of similar soil chemical, physical, and microbial properties, especially in locations with established legacy effects (in our case, more than 20 years of a brome grass monoculture). Our findings support and add to new information regarding temporal changes in plant-climate-soil interactions which have not been conducted previously for pasture sites dominated by cool-season grasses such as brome grass over several decades of development.

Acknowledgements

The authors acknowledge an internal funding grant provided by the University of Nebraska-Lincoln's Research and Development Office and Agricultural Research Service (ARS), USDA. We acknowledge Les Howard and Paul Koerner for their valuable technical assistance.

Abbreviations

pH	potential of hydrogen
N	nitrogen
ECa	apparent electroconductivity
OC	organic carbon
WDRF BpH	Woodruff buffer pH

Author details

Taity Changa[1], Jane Asiyo Okalebo[1]* and Shaokun Wang[2]

1 School of Natural Resources, University of Nebraska, Lincoln, NE, USA

2 Urat Desert-Grassland Research Station, Northwest Institute of Eco-Environment and Resources, Chinese Academy of Sciences, Lanzhou, China

*Address all correspondence to: jane.okalebo@gmail.com

References

[1] Kulmatiski A, Beard KH, Stevens JR, Cobbold SM. Plant-soil feedbacks: A meta-analytical review. Ecology Letters. 2008;**11**(9):980-992. DOI: 10.1111/j.1461-0248.2008.01209.x

[2] Ludwig M, Wilmes P, Schrader S. Measuring soil sustainability via soil resilience. The Science of the Total Environment. 2018;**626**:1484-1493. DOI: 10.1016/j.scitotenv.2017.10.043

[3] Bardgett RD, van der Putten WH. Belowground biodiversity and ecosystem functioning. Nature. 2014;**515**(7528):505-511. DOI: 10.1038/nature13855

[4] Sergaki C, Lagunas B, Lidbury I, Gifford ML, Schäfer P. Challenges and approaches in microbiome research: From fundamental to applied. Frontiers in Plant Science. 2018;**9**:1205. DOI: 10.3389/fpls.2018.01205

[5] Tu J, Qiao J, Zhu Z, Li P, Wu L. Soil bacterial community responses to long-term fertilizer treatments in paulownia plantations in subtropical China. Applied Soil Ecology. 2018;**124**:317-326. DOI: 10.1016/j.apsoil.2017.09.036

[6] Hazard C, Johnson D. Does genotypic and species diversity of mycorrhizal plants and fungi affect ecosystem function? The New Phytologist. 2018;**220**(4):1122-1128. DOI: 10.1111/nph.15010

[7] Schloter M, Dilly O, Munch JC. Indicators for evaluating soil quality. Agriculture, Ecosystems & Environment. 2003;**98**(1-3):255-262. DOI: 10.1016/S0167-8809(03)00085-9

[8] Liang ZB, Lee DJ, Dweikat IM, Wedin DA, Yuen GY, Drijber RA. Molecular diversity of Arbuscular Mycorrhizae in roots of Juniperus virginiana invasive to grasslands. Soil Science Society of America Journal. 2017;**81**(3):526-536. DOI: 10.2136/sssaj2016.05.0133

[9] Zhou Y, Boutton TW, Wu XB, McCulley R. Soil carbon response to woody plant encroachment: Importance of spatial heterogeneity and deep soil storage. Journal of Ecology. 2017;**105**(6):1738-1749. DOI: 10.1111/1365-2745.12770

[10] Wang J-F, Stein A, Gao B-B, Ge Y. A review of spatial sampling. Spatial Statistics. 2012;**2**:1-14. DOI: 10.1016/j.spasta.2012.08.001

[11] Doll EH. Gross DL. EC119 Brome Grass for Erosion Control. Historical Materials from University of Nebraska-Lincoln Extension [Online]. 1938. Available from: https://digitalcommons.unl.edu/extensionhist/2096

[12] Stubbendieck J, Hatch SL, Kjar KJ. North American Range Plants. 2nd ed. Lincoln Neb., London: University of Nebraska Press; 1982

[13] Piper CL, Lamb EG, Siciliano SD. Smooth brome changes gross soil nitrogen cycling processes during invasion of a rough fescue grassland. Plant Ecology. 2015;**216**(2):235-246. DOI: 10.1007/s11258-014-0431-y

[14] Snell LK, Guretzky JA, Jin VL, Drijber RA, Mamo M. Ruminant urine increases uptake but decreases relative recovery of nitrogen by smooth bromegrass. Crop, Forage & Turfgrass Management. 2017;**3**(1):cftm2016.03.0022. DOI: 10.2134/cftm2016.03.0022

[15] Piper CL, Siciliano SD, Winsley T, Lamb EG, de Deyn G. Smooth brome invasion increases rare soil bacterial species prevalence, bacterial species

richness and evenness. Journal of Ecology. 2015;**103**(2):386-396. DOI: 10.1111/1365-2745.12356

[16] Grigera MS, Drijber RA, Eskridge KM, Wienhold BJ. Soil microbial biomass relationships with organic matter fractions in a Nebraska corn field mapped using apparent electrical conductivity. Soil Science Society of America Journal. 2006;**70**(5):1480-1488. DOI: 10.2136/sssaj2005.0331

[17] Lauber CL, Ramirez KS, Aanderud Z, Lennon J, Fierer N. Temporal variability in soil microbial communities across land-use types. The ISME Journal. 2013;7(8):1641-1650. DOI: 10.1038/ismej.2013.50

[18] e Silva MCP, Dias ACF, van Elsas JD, Salles JF. Spatial and temporal variation of archaeal, bacterial and fungal communities in agricultural soils. PLoS One. 2012;7(12):e51554. DOI: 10.1371/journal.pone.0051554

[19] Segal LM et al. Bacterial and archaeal ammonia oxidizers respond differently to long-term tillage and fertilizer management at a continuous maize site. Soil and Tillage Research. 2017;**168**:110-117. DOI: 10.1016/j.still.2016.12.014

[20] Nakajima T, Shrestha RK, Lal R. On-farm assessments of soil quality in Ohio and Michigan. Soil Science Society of America Journal. 2016;**80**(4):1020-1026. DOI: 10.2136/sssaj2016.01.0003

[21] O'Brien SL et al. Spatial scale drives patterns in soil bacterial diversity. Environmental Microbiology. 2016;**18**(6):2039-2051. DOI: 10.1111/1462-2920.13231

[22] Bossio DA, Scow KM, Gunapala N, Graham KJ. Determinants of soil microbial communities: Effects of agricultural management, season, and soil type on phospholipid fatty acid profiles. Microbial Ecology. 1998;**36**(1):1-12. DOI: 10.1007/s002489900087

[23] Bartram AK et al. Exploring links between pH and bacterial community composition in soils from the Craibstone experimental farm. FEMS Microbiology Ecology. 2014;**87**(2):403-415. DOI: 10.1111/1574-6941.12231

[24] Broughton LC, Gross KL. Patterns of diversity in plant and soil microbial communities along a productivity gradient in a Michigan old-field. Oecologia. 2000;**125**(3):420-427. DOI: 10.1007/s004420000456

[25] Rousk J et al. Soil bacterial and fungal communities across a pH gradient in an arable soil. The ISME Journal. 2010;**4**(10):1340-1351. DOI: 10.1038/ismej.2010.58

[26] Naveed M et al. Spatial variability of microbial richness and diversity and relationships with soil organic carbon, texture and structure across an agricultural field. Applied Soil Ecology. 2016;**103**:44-55. DOI: 10.1016/j.apsoil.2016.03.004

[27] Peralta NR, Costa JL. Delineation of management zones with soil apparent electrical conductivity to improve nutrient management. Computers and Electronics in Agriculture. 2013;**99**:218-226. DOI: 10.1016/j.compag.2013.09.014

[28] Triantafilis J, Lesch SM. Mapping clay content variation using electromagnetic induction techniques. Computers and Electronics in Agriculture. 2005;**46**(1-3):203-237. DOI: 10.1016/j.compag.2004.11.006

[29] Miura T, Makoto K, Niwa S, Kaneko N, Sakamoto K. Comparison

of fatty acid methyl ester methods for characterization of microbial communities in forest and arable soil: Phospholipid fraction (PLFA) versus total ester linked fatty acids (EL-FAME). Pedobiologia. 2017;**63**:14-18. DOI: 10.1016/j.pedobi.2017.04.002

[30] Schutter ME, Dick RP. Comparison of fatty acid methyl ester (FAME) methods for characterizing microbial communities. Soil Science Society of America Journal. 2000;**64**(5):1659-1668. DOI: 10.2136/sssaj2000.6451659x

[31] Hendrickson JR, Wienhold BJ, Berdahl JD. Decomposition rates of native and improved cultivars of grasses in the northern great plains. Arid Land Research and Management. 2010;**15**(4):347-357. DOI: 10.1080/153249801753127633

[32] Vinton MA, Goergen EM. Plant–soil feedbacks contribute to the persistence of *Bromus inermis* in Tallgrass prairie. Ecosystems. 2006;**9**(6):967-976. DOI: 10.1007/s10021-005-0107-5

[33] Greenquist MA et al. Effects of nitrogen fertilization and dried distillers grains supplementation: Forage use and performance of yearling steers. Journal of Animal Science. 2009;**87**(11):3639-3646. DOI: 10.2527/jas.2009-1795

[34] Lesch SM, Strauss DJ, Rhoades JD. Spatial prediction of soil salinity using electromagnetic induction techniques: 2. An efficient spatial sampling algorithm suitable for multiple linear regression model identification and estimation. Water Resources Research. 1995;**31**(2):387-398. DOI: 10.1029/94WR02180

[35] Smith JL, Doran JW. Measurement and use of pH and electrical conductivity for soil quality analysis. In: Doran JW, Jones AJ, editors. SSSA Special Publications, Methods for Assessing Soil Quality. Madison, WI, USA: Soil Science Society of America; 1997. pp. 169-185

[36] Bray RH, Kurtz LT. Determination of total, organic, and available forms of phosphorus in soils. Soil Science. 1945;**59**(1):39-46

[37] Murphy J, Riley JP. A modified single solution method for the determination of phosphate in natural waters. Analytica Chimica Acta. 1962;**27**:31-36. DOI: 10.1016/S0003-2670(00)88444-5

[38] Mulvaney RL. Nitrogen-inorganic forms. Methods of Soil Analysis: Part 3 Chemical Methods. Wiley Online Library. 1996;**5**:1123-1184

[39] Rhoades JD, Manteghi NA, Shouse PJ, Alves WJ. Estimating soil salinity from saturated soil-paste electrical conductivity. Soil Science Society of America Journal. 1989;**53**(2):428-433

[40] Kroppenstedt RM, Goodfellow M, Minnikin DE. Chemical Methods in Bacterial Systematics. London: Academic Press; 1985

[41] Stahl PD, Klug MJ. Characterization and differentiation of filamentous fungi based on fatty acid composition. Applied and Environmental Microbiology. 1996;**62**(11):4136-4146

[42] Ringelberg DB, Stair JO, Almeida J, Norby RJ, O'Neill EG, White DC. Consequences of rising atmospheric carbon dioxide levels for the belowground microbiota associated with white oak. Journal of Environmental Quality. 1997;**26**(2):495-503. DOI: 10.2134/jeq1997.00472425002600020022x

[43] Olsson PA, Johansen A. Lipid and fatty acid composition of hyphae and spores of arbuscular mycorrhizal fungi at different growth stages. Mycological Research. 2000;**104**(4):429-434. DOI: 10.1017/S0953756299001410

[44] Dray S, Dufour A-B. The ade4 package: Implementing the duality diagram for ecologists. Journal of Statistical Software. 2007;**22**(4):1-20. DOI: 10.18637/jss.v022.i04

[45] Kassambara A. Mundt F. Factoextra: Extract and visualize the results of multivariate data analyses. 2015. Available from: https://cran.r-project.org/web/packages/factoextra/index.htm [Accessed: 03 March 2019]

[46] Eigenberg RA, Doran JW, Nienaber JA, Ferguson RB, Woodbury BL. Electrical conductivity monitoring of soil condition and available N with animal manure and a cover crop. Agriculture, Ecosystems & Environment. 2002;**88**(2):183-193. DOI: 10.1016/S0167-8809(01)00256-0

[47] Wood M. Soil Biology. Glasgow, New York: Blackie; 1989

[48] Strickland MS, Rousk J. Considering fungal:bacterial dominance in soils – Methods, controls, and ecosystem implications. Soil Biology and Biochemistry. 2010;**42**(9):1385-1395. DOI: 10.1016/j.soilbio.2010.05.007

[49] Grigera M, Drijber R, Wienhold B. Redistribution of crop residues during row cultivation creates a biologically enhanced environment for soil microorganisms. Soil and Tillage Research. 2007;**94**(2):550-554. DOI: 10.1016/j.still.2006.08.016

[50] Šťovíček A, Azatyan A, Soares MIM, Gillor O. The impact of hydration and temperature on bacterial diversity in arid soil mesocosms. Frontiers in Microbiology. 2017;**8**:1078. DOI: 10.3389/fmicb.2017.01078

[51] Bååth E, Anderson T-H. Comparison of soil fungal/bacterial ratios in a pH gradient using physiological and PLFA-based techniques. Soil Biology and Biochemistry. 2003;**35**(7):955-963. DOI: 10.1016/S0038-0717(03)00154-8

[52] Kaiser C et al. Belowground carbon allocation by trees drives seasonal patterns of extracellular enzyme activities by altering microbial community composition in a beech forest soil. The New Phytologist. 2010;**187**(3):843-858. DOI: 10.1111/j.1469-8137.2010.03321.x

[53] Moore-Kucera J, Dick RP. PLFA profiling of microbial community structure and seasonal shifts in soils of a Douglas-fir chronosequence. Microbial Ecology. 2008;**55**(3):500-511. DOI: 10.1007/s00248-007-9295-1

[54] Santalahti M, Sun H, Jumpponen A, Pennanen T, Heinonsalo J. Vertical and seasonal dynamics of fungal communities in boreal scots pine forest soil. FEMS Microbiology Ecology. 2016;**92**(11):1-12. ISSN: 0168-6496. DOI: 10.1093/femsec/fiw170

[55] Armstrong A et al. Temporal dynamics of hot desert microbial communities reveal structural and functional responses to water input. Scientific Reports. 2016;**6**:34434. DOI: 10.1038/srep34434

[56] Papatheodorou EM, Argyropoulou MD, Stamou GP. The effects of large- and small-scale differences in soil temperature and moisture on bacterial functional diversity and the community of bacterivorous nematodes. Applied Soil Ecology. 2004;**25**(1):37-49. DOI: 10.1016/S0929-1393(03)00100-8

[57] Manzoni S, Taylor P, Richter A, Porporato A, Ågren GI. Environmental and stoichiometric controls on microbial carbon-use efficiency in soils. New Phytologist. 2012;**196**(1):79-91. DOI: 10.1111/j.1469-8137.2012.04225.x

[58] Semchenko M et al. Fungal diversity regulates plant-soil feedbacks in temperate grassland. Science Advances. 2018;**4**(11):eaau4578. DOI: 10.1126/sciadv.aau4578

[59] Viaud V et al. Landscape-scale analysis of cropping system effects on soil quality in a context of crop-livestock farming. Agriculture, Ecosystems & Environment. 2018;**265**:166-177. DOI: 10.1016/j.agee.2018.06.018

[60] Gosling P, Hodge A, Goodlass G, Bending GD. Arbuscular mycorrhizal fungi and organic farming. Agriculture, Ecosystems & Environment. 2006;**113**(1-4):17-35. DOI: 10.1016/j.agee.2005.09.009

[61] Johnson NC, Wilson GWT, Bowker MA, Wilson JA, Miller RM. Resource limitation is a driver of local adaptation in mycorrhizal symbioses. Proceedings of the National Academy of Sciences of the United States of America. 2010;**107**(5):2093-2098. DOI: 10.1073/pnas.0906710107

[62] Fink KA, Wilson SD. *Bromus inermis* invasion of a native grassland: Diversity and resource reduction. Botany. 2011;**89**(3):157-164

[63] Bardgett RD, Mawdsley JL, Edwards S, Hobbs PJ, Rodwell JS, Davies WJ. Plant species and nitrogen effects on soil biological properties of temperate upland grasslands. Functional Ecology. 1999;**13**(5):650-660. DOI: 10.1046/j.1365-2435.1999.00362.x

[64] Grman E. Plant species differ in their ability to reduce allocation to non-beneficial arbuscular mycorrhizal fungi. Ecology. 2012;**93**(4):711-718. DOI: 10.1890/11-1358.1

[65] Broeckling CD, Broz AK, Bergelson J, Manter DK, Vivanco JM. Root exudates regulate soil fungal community composition and diversity. Applied and Environmental Microbiology. 2008;**74**(3):738-744. DOI: 10.1128/AEM.02188-07

[66] Gollotte A, van Tuinen D, Atkinson D. Diversity of arbuscular mycorrhizal fungi colonising roots of the grass species Agrostis capillaris and Lolium perenne in a field experiment. Mycorrhiza. 2004;**14**(2):111-117. DOI: 10.1007/s00572-003-0244-7

[67] Higo M, Isobe K, Miyazawa Y, Matsuda Y, Drijber RA, Torigoe Y. Molecular diversity and distribution of indigenous arbuscular mycorrhizal communities colonizing roots of two different winter cover crops in response to their root proliferation. Korean Journal of Microbiology. 2016;**54**(2):86-97. DOI: 10.1007/s12275-016-5379-2

[68] Mamet SD, Lamb EG, Piper CL, Winsley T, Siciliano SD. Archaea and bacteria mediate the effects of native species root loss on fungi during plant invasion. The ISME Journal. 2017;**11**(5):1261-1275. DOI: 10.1038/ismej.2016.205

[69] Bru D et al. Determinants of the distribution of nitrogen-cycling microbial communities at the landscape scale. The ISME Journal. 2011;**5**(3):532-542. DOI: 10.1038/ismej.2010.130

[70] Wymore AS et al. The influence of time and plant species on the composition of the decomposing

bacterial community in a stream ecosystem. Microbial Ecology. 2016;**71**(4):825-834. DOI: 10.1007/s00248-016-0735-7

[71] Maróti G, Kondorosi E. Nitrogen-fixing rhizobium-legume symbiosis: Are polyploidy and host peptide-governed symbiont differentiation general principles of endosymbiosis? Frontiers in Microbiology. 2014;**5**:326. DOI: 10.3389/fmicb.2014.00326

Chapter 13

Beetles and Meteorological Conditions: A Case Study

Marcos Paulo Gomes Gonçalves

Abstract

The meteorological factors study in the beetle population dynamics, as well as its association with vegetation, is of fundamental importance for understanding the variation that occurs in its population. Thus, it was reported the influence of temperature, humidity, insolation and precipitation on the beetles in general and it was presented a case study that examined the relationship between time and population fluctuation of curculionids in Mata de Cocal and an area used for crop rotation and animal grazing, in the city of Teresina, Brazil, from August 2011 to July 2012. It was verified that beetles populations certain are governed and conditioned by meteorological variables to a greater or lesser extent depending on the characteristics of the community itself and the biotic and abiotic environmental factors of the area where they live: the temperature that changes the its metabolic rate, the insolation and humidity that can affect its fertility and longevity can be cited as examples. From the case presented, It was found that the Curculionidae community has a positive association with precipitation and humidity and a negative association with insolation and temperature, being that in native forests curculionids are not as dependent on meteorological variables as in agricultural fields.

Keywords: ecology, biometeorology, entomology, Coleoptera, Curculionidae

1. Introduction

Coleopteran insects, which they are popularly known as beetles, are important indicators of the environment quality, as they are susceptible to climatic variations and occupy a habitats diversity [1–3]. Among the beetles, the family Curculionidae Latreille, 1802, is a very important group present in forest ecosystems due to the role of wood deterioration they make.[1]

Insects like all living beings are subject to the nature forces, so there may be an influence that can inhibit or favor the species certain development. For example, the hottest and driest events on El Niño are having an alarming effect on biodiversity in the Amazon Rainforest, contributing to the reduction of insects in the Amazon and across the globe[2] [4]. Intense droughts and forest fires during the last El Niño climatic phenomenon, combined with human disturbances - deforestation and

[1] Curculionids are known as *Rüsselkäfer* or *Rüssler* in German, *charançon* in French and weevil in English. In Brazil, they are known as *gorgulhos*, *carneirinhos* or *bicudos* depending on the region.

[2] El Niño is an atmospheric-oceanic event is the abnormal warming of surface waters of the tropical Pacific Ocean. It affects regional and global meteorological conditions, changing wind dynamics around the world, changing rainfall patterns in tropical and mid-latitudes.

burning to clear areas for agricultural production and the sale of illegal wood - led to the reduction of beetles in the Amazon, the numbers dropped in half - with effects that can last at least two years, according to the researchers: about 8000 beetles were counted in 2010; 3700 in 2016, after El Niño; and 2600 beetles in 2017 [4].

Thus, in view of the climate changes that are already occurring and that are predicted for the coming years, the meteorological factors study in the beetle population dynamics, as well as its association with vegetation, is of fundamental importance for understanding the variation that occurs in its population.

In this way will be discussed the influence of temperature, humidity, insolation and precipitation on the beetles in general, firstly, in this chapter. And a second moment, as it is a relevant subject and to complement the scientific literature about the family Curculionidae when associated with the native forest and the areas agriculture and meteorological variables, a case study will be reported that aimed to analyze the curculionids relationship to temporal factors (precipitation, insolation, temperature and humidity) in Mata de Cocal (native forest) and in an area used for crop rotation and animal grazing (agricultural field).

2. Beetles and meteorological conditions

Around the world, extreme climatic events increase and expose the vulnerability of insect populations, altering the biological functioning and the community structure of these invertebrates. Depending on a given meteorological situation, in addition to affecting the forests biodiversity and the ecosystem functioning, the community of a given species can become a pest for an agricultural crop.[3]

Thus, beetles are subject to natural forces that can inhibit or favor the species certain development, because they are associated with the relative humidity, precipitation, insolation and temperature mainly [1]. The following will be seen how each of these factors can influence the insects.

The ambient temperature regulates the insects' internal temperature, as they are pecilothermal animals - they do not have a thermoregulation system. Temperature gives insects a change in their metabolic rate interfering with their development – egg, larva and pupa phase – and their behavior [2, 3]. Thus, when the ambient temperature is favorable, the smaller insects benefit from the easy heat exchange with the environment, also having more efficient respiratory and circulatory activity, more intense metabolic activity and greater capacity to use food resources [2]. In general, for insects, the optimum temperature for the fastest development and the offspring largest number is close to 25° C and the optimum range for the insects most is between 15 and 38° C [7].

Unfavorable temperatures decrease the insect's metabolic activity. Very low temperatures can lead to an almost absolute killing of certain species of beetles [3]. For example, −16° C is the critical value for the North American tree bark beetle species *Dendroctonus frontalis* Zimm, 1868, which belongs to the Curculionidae family [3, 8]. There is also a negative influence of the average temperature increase in the forest environment in relation to the European beetle *Ips typographus* (L.), 1758, which also belongs to the same family [3, 9–10].

[3] In South America, weather conditions (temperature, precipitation and wind dynamics) are related to the displacement route - which involves Paraguay, Argentina, Brazil and Uruguay - of the locust cloud that can destroy a plantation in just one day [5]. Too, the increase in temperature relates positively to the generations number of the fruit fly, *Anastrepha grandis* [6]. Consequently, high temperatures, in addition to increasing the population of the fruit fly, favor the spread of this insect in areas that do not yet inhabit and may thus accentuate losses for agricultural production [6].

Some insects, like plants, need a certain number of sunshine hours to complete their development. These hours interfere with fertility and longevity [2]. In addition to humidity, precipitation, temperature, insolation is decisive for the distribution the insects fauna richness [1, 11, 12]. Some beetles families, in northeastern Brazil, have associated this meteorological variable positively [1, 11, 13].

The amount of water contained in the insects body varies between 70 to 90%, which depends on the food type they consume and the environment in which they live [2]. The relative humidity is directly related to the insects exposure or their protection, the low moisture content can affect the physiology, longevity, development and insects oviposition [14]. For most insects, the favorable range for greater longevity, fertility and development speed ranging from 40 to 80% [7].

The rainfall may indirectly interfere with the insects. In study, in the Ponta Grossa city- Brazil, in areas with different anthropization degrees and plant succession, it was found that the insects seasonality is related to the rainy season, as there is a greater availability of food resources, which provides a high population peak [15]. However, the excess water can be harmful to plants, it can cause water stress and soil get soggy [1].

There are brazilian studies in which there was the highest record of specimens sampled in the rainy season. For example, in a fragment of Atlantic Forest, in the Gurjaú Ecological Reserve, Santos obtained 71.5% of the total sampled specimens in the rainy season (May to July 2003) and 28.5% of the total sampled beetles in the dry season (October and November 2003) [16].

Few species of coprophagous beetles survive in areas with an average annual rainfall of less than 250 mm, the individuals number increases only at the beginning of the first rains and with the rise in air temperature [17]. Corroborating this understanding, studies have shown that, in pastures, the scarab community population fluctuation is predominantly governed by the region's macro-climatic variables and that the beetles presence is seasonal and conditioned by temporal parameters, correlating to the compensated average air temperature negatively and to the compensated average relative humidity and the pluviometric precipitation positively [18, 19]. In Mata de Cocal, it was confirmed canonical correlation between sets of variables monthly record of beetles occurrence and monthly weather data [1].

To attest to the meteorological factors influence of on the curculionid community population fluctuation, in an area of forest native (Mata de Cocal) and agriculture (crop rotation with grazing animals), a case study will be presented in the next section.

3. Case study

3.1 Methodology

This study was carried out at the Agricultural Sciences Center localized at the Federal University of Piauí (UFPI), in the Teresina city, from August 2011 to July 2012.

Teresina's climate, based on the classification made by Thornthwaite and Mather, is C1sA'a', it is characterized as dry subhumid, megathermic, with moderate hydrous surplus in summer and a potential evapotranspiration of 32.2% in the trimester – September, October and November [20].

A native vegetation area was used for sampling, its respective latitude and longitude is 5° 2'52 "S and 42 ° 47'11" W, which it presents formation denominated Mata de Cocal. The other area used is characterized by being an agricultural field with respective latitude and longitude is 52° 2 '57' 'S and 42° 46' 57 "W, where during the present study there was a concomitant and rotational cultivation of pigeon pea (*Cajanus cajan* (L.) Millsp.), cowpea (*Vigna unguiculata* (L.) Walp), corn (*Zea mays L.)* and watermelon *(Citrullus lanatus* (Thunb.) Matsum & Nakai), in addition to the occasional presence of sheep *(Ovis aries* L., 1758*)* and goats *(Capra aegagrus hircus* L., 1758).

Geographically, this native vegetation is concentrated in the Piauí and Maranhão states, most western portion of the Northeast region, widely occupied by dicots and palm trees formations [21]. In vegetation terms, an extensive mosaic is formed with so different physiognomies that the region's specialty in having disseminated features with different species and structures is observed in little space, accompanied by alterion in soil and climate [22]. The Piauí and Maranhão are an extensive ecotonal zone located between the Amazon sub-humid and the Northeastern semi-arid, transition region [23].

For the samples, unattractive pitfall traps were used to capture the insects. (These traps were also used, for example, see [19, 24])Three independent sample units are used, called "stations" located at ground level. Each station consisted of four plastic containers with a capacity of 500 mL, a diameter of 10 cm and a height of 11 cm, covered by a plastic plate 20 cm in diameter, suspended by pieces of wood approximately 15 cm, to prevent or reduce water entry in the rainy season. The containers were buried with the edge at ground level, with the addition of 200 ml of 4% formaldehyde solution as fixative liquid. The flasks were connected by metal bulkheads, 100 cm long by 20 cm high, to direct the insects to the station's containers. The stations were arranged in the shape of a Y. **Figure 1** (adapted) shows the schematic drawing of the station.

Three identical stations were randomly arranged in each area. Each station kept a distance of 5 m between them. They were obtained weekly samples, totaling 52 samples from August 2011 to July 2012. At the UFPI's entomology laboratory, screening and identification of Coleoptera specimens was carried out, besides the deposit of the material sampled for its preservation, for possible future studies.

The monthly meteorological data of the compensated average air temperature, compensated relative humidity, precipitation and insolation were obtained through the network of the National Institute of Meteorology – INMET. The weather station from which the weather data originated has a respective latitude and longitude of 5 ° 220.16 "S and 42 ° 48'14.77"W and an altitude of 75 m. The areas of native forest and agricultural field in which the samples were obtained are within a radius of 3 km from the aforementioned weather station.

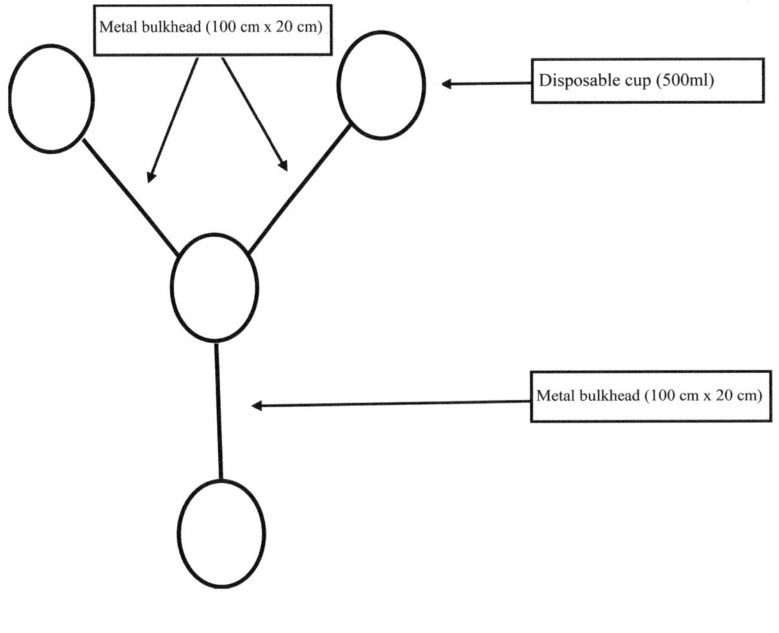

Figure 1.
Station design (adapted).

Faunistic indices were calculated: dominance, abundance, frequency and constancy, as well as diversity indices, H variance and confidence interval with the ANAFAU software [25]. Correlations were performed using the Pearson method between the records of monthly occurrence of the population fluctuation of the subfamilies and the monthly data of the meteorological variables using the software BIOESTAT version 5.3 [26]. Later, the canonical correlation was performed between the sets of the monthly occurrence of the population fluctuation of the subfamilies and the set of monthly meteorological data by the same software.

3.2 Results and discussion

3.2.1 Population fluctuation and fauna measures

Table 1 shows that in the period analyzed in the Mata de Cocal 591 insects of the Curculionidae family were sampled and distributed in 2 subfamilies: Molytinae Schönherr, 1823, and Scolytinae Latreille, 1804, with 22 and 569 specimens respectively. Already in the agricultural field 63 insects of this family were collected and distributed in Molytinae and Scolytinae too, with 41 and 22 specimens respectively. In total, 654 Curculionidae beetles were collected, 90.37% in the Mata de Cocal area and 9.63% in the agriculture area.

The subfamily Molytinae was more numerous in the agricultural field than the same subfamily in Mata de Cocal, while Scolytinae presented more specimens in this area than in that one.

The population fluctuation of beetles ranged from a minimum of 0 specimens in Mata de Cocal, in the months of August, October and November 2011, and March and July 2012, to a maximum of 513 in April 2012, which represented 86.80% of the insects collected in that area. In the agricultural field, in the months of September, October and November 2011, no specimen was collected from the family studied, while 32 were collected in the month of April 2012, which represented 50.79% of the specimens collected in that area. For all subfamilies, the population peak occurred in April.

The population peak of a given insect community occurs when the set of homeostatic mechanisms in nature, in relation to one's own family or species, as intrinsics – internal – and extrinsics– external–, is favorable to its development in a given habitat, over a given period of time [1]. Intrinsic mechanisms depend on the members of the population themselves, as is the case with intraspecific competition, while extrinsic mechanisms depend on the participation of something outside the population itself, such as interspecific competition, food and space restrictions, weathering, parasitism, predatism and meteorological variations [1]. Thus, the April month had the most favorable environmental homeostatic conditions for the studied communities of curculionids.

Table 2 shows that all subfamilies were dominant, constant and accessory. As for abundance, subfamilies were very abundant in Mata de Cocal, while they were common in the agricultural field. Also, the fauna indexes – Shannon-Weaner diversity, Margalef wealth and equitability – were different among the curculionid communities of the studied areas. The highest values of these indexes were obtained in the agriculture area.

Normally the difference between insect community is the relationship of insects to biotic factors – vegetables diversity; intra and interspecific competitions; mammal diversity– and abiotic factors – such as seasonality with its characteristics of temperature, air humidity, rainfall and photoperiod; soil edaphic conditions; among other factors [1]. In this way, the association of the Curculionidae community to the biotic and abiotic factors of which they are inserted may explain the difference in terms of abundance and fauna indexes presented in the studied areas.

Area	Subfamily	2011					2012							Total
		Aug.	Sep.	Oct.	Nov.	Dec.	Jan.	Feb.	Mar.	Apr.	May	Jun.	Jul.	
Mata de Cocal	Scolytinae	0	0	0	0	0	0	0	0	506	45	18	0	569
	Molytinae	0	1	0	0	1	8	4	0	7	0	1	0	22
Agricultural field	Scolytinae	0	0	0	0	0	0	0	0	20	1	1	0	22
	Molytinae	1	0	0	0	3	8	7	8	12	0	0	2	41
Total		1	1	0	0	4	16	11	8	545	46	20	2	654

Table 1.
Monthly records of population fluctuation of curculionids data collected in the Mata de Cocal and in the agricultural field.

Area	Subfamily	Dominance	Abundance	Frequency	Constancy
Mata de cocal	Scolytinae	D*	va	F	Y
	Molytinae	D	va	F	Y
Agricultural field	Scolytinae	D	c	F	Y
	Molytinae	D	c	F	Y

**D = dominant; F = frequent; Y = accessory; c = common; va = very abundant; a =abundant.*
***Mata de cocal:** Diversity Index (Shannon-Weaner) => H = 0.1590 H Confidence Interval (P = 0.05) => [0.156938; 0.161106] Wealth Index (Margalef) => ALFA = 0.1567 Uniformity or Equitability Index => E = 0.2294.*
***Agricultural field:** Diversity Index (Shannon-Weaner) => H = 0.6470 H Confidence Interval (P = 0.05) => [0.637532; 0.656375] Wealth Index (Margalef) => ALFA = 0.2414 Uniformity or Equitability Index => E = 0.9334.*

Table 2.
Faunistic analysis of the curculionid subfamilies sampled in Mata de Cocal and in the agricultural field.

3.2.2 Curculionids and meteorological variables

3.2.2.1 Weather data

Throughout the study, there was an average insolation of 8.4 h.d-1, an average relative humidity of 72.4%, an average daily compensated air temperature of 27.4°C and 1218.9 mm of rainfall. When these values are compared to the climatological normal from 1980 to 2009 calculated by Bastos and Andrade Júnior – who it had an average insolation of 7.8 h.d-1, relative humidity of 69.8%, average air temperature of 28.2°C and 1351.9 mm annual rainfall–, it was verified that the insolation and humidity were higher while precipitation and temperature were below [27]. The record of monthly meteorological data for the period is shown in **Table 3**.

Year	Month	Insolation (h.d^{-1})	Relative humidity (%)	Precipitation (mm)	Average temperature (° C)
2011	Aug.	10.1	65.4	10.8	27.2
	Sep.	10.3	58.1	0.6	28.5
	Oct.	8.8	66.1	167.6	28.6
	Nov.	7.8	74.8	124.6	27.4
	Dec.	8.6	68.2	23.4	28.3
2012	Jan.	6.4	75.7	133.1	27.1
	Feb.	5.6	83.4	317.1	26.2
	Mar.	7.4	85	264	26.3
	Apr.	7.7	82.5	121	26.7
	May	9.3	77.3	31.3	27.2
	Jun.	8.8	72.7	25.4	27.2
	Jul.	10	60.8	0	27.5

Source: INMET Network data.

Table 3.
Meteorological data (insolation, compensated average relative humidity, precipitation and average compensated air temperature) from August 1, 2011 to July 31, 2012 in Teresina.

3.2.2.2 Relationship between curculionids and meteorological variables

Beetles like all living things are subject to the nature forces, so there is a relevant influence in a way that can inhibit or favor a particular group of insects. Studying the relationship between meteorological factors and the population fluctuation of beetles is important for understanding the variation that occurs in their population.

For investigation and understanding of this association in the Mata de Cocal and in the agricultural field, the monthly records of the subfamilies fluctuation and the monthly meteorological data were correlated.

Significantly, in Mata de Cocal, the subfamily Molytinae was negatively associated with heat stroke (−0.61; $p < 0.05$), while the same subfamily, in an agricultural area, was negatively associated with insolation (−0.70; $p < 0.05$) and temperature (−0.61; $p < 0.05$), and positively with humidity (0.68; $p < 0.05$) and precipitation (0.56; $p < 0.10$). In both areas, the Scolytinae subfamily had no significant association with a p-value equal to>0.10. **Figures 2–5** show that only the subfamily Molytinae, in the agriculture area, was significantly associated with all meteorological variables (**Table 4**).

Canonical correlation was performed between the monthly records of the subfamilies and the data set related to insolation, humidity, precipitation and temperature.

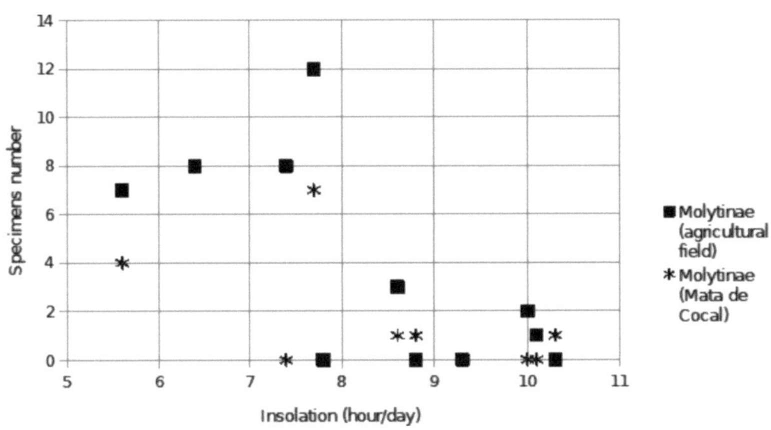

Figure 2.
Dispersion diagram: the daily insolation versus the monthly population fluctuation of Molytinae.

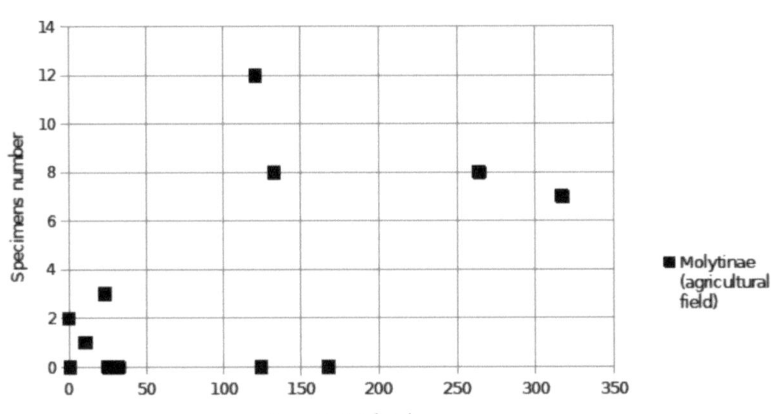

Figure 3.
Dispersion diagram: the precipitation data versus the monthly population fluctuation of Molytinae.

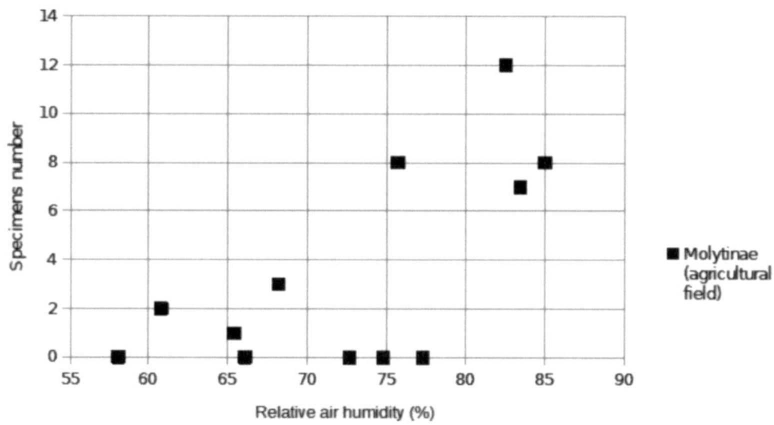

Figure 4.
Dispersion diagram: the monthly relative air humidity data versus the monthly population fluctuation of Molytinae.

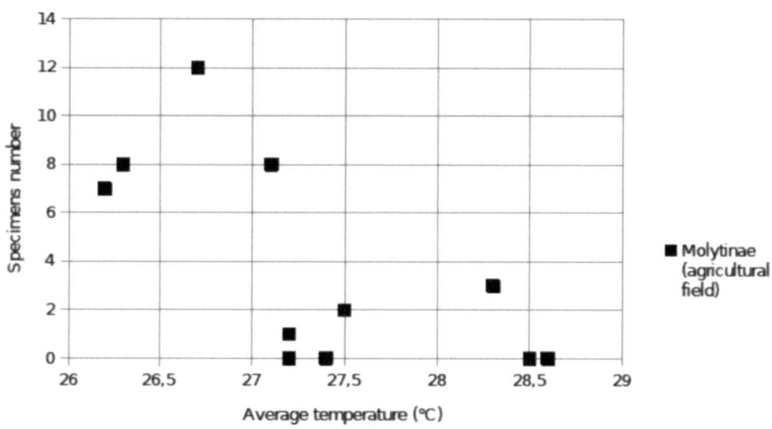

Figure 5.
Dispersion diagram: the monthly average temperature data versus the monthly population fluctuation of Molytinae.

Subfamily Molytinae	Insolation ($h.d^{-1}$)		Humidity (%)		Precipitation (mm)		Temperature (°C)	
	(r)	p- value	(r)	p- value	(r)	p- value	(r)	p- value
Mata de cocal	−0.61	< 0.05	0.42	ns*	0.29	ns	−0.38	ns
Agricultural field	−0.63	< 0.05	0.68	< 0.05	0.56	< 0.10	−0.70	< 0.05

N (pairs) = 12; degrees of freedom = 10;
** ns = not significant.*

Table 4.
Correlation between monthly records of population curculionids fluctuation and monthly meteorological data in Mata de Cocal and in agriculture area.

Canonical correlations and eigenvalues were observed in **Tables 5** and **6**. The correlation between the pair of canonical variables, also called canonical R, was approximately 0.79 in both areas, which represents the best possible correlation between any linear combination of the monthly meteorological data with the records monthly of the subfamilies Curculionidae. The Canonical R statistic expresses the relationship magnitude between the variables sets.

	Mata de cocal		Agricultural field	
	Molytinae	Scolytinae	Molytinae	Scolytinae
V	0.93	0.21	0.97	0.40

Mata de cocal: *canonical R= 0.79; canonical R^2= 0.63; χ2 = 11.5400, 15; Degrees of freedom= 8; p-value < 0.1729.*
Agricultural field: *canonical R= 0.79; canonical R^2 = 0,62; χ2 = 10.6400, 15; Degrees of freedom= 8; p-value <0.2229.*

Table 5.
Canonical correlation between the beetle population fluctuation records and the meteorological data set (V).

	Insolation (h.d^{-1})	Relative air humidity (%)	Precipitation (mm)	Average temperature (° C)
U (mata de cocal)	−0.80	0.42	0.42	−0.33
U (Agricultural field)	−0.87	0.86	0.82	−0.87

Mata de cocal: *canonical R= 0.79; canonical R^2= 0.63; χ2 = 11.5400, 15; Degrees of freedom= 8; p-value < 0.1729.*
Agricultural field: *canonical R = 0,79; canonical R^2= 0.62; χ2 = 10.6400, 15; Degrees of freedom= 8; p-value < 0.2229.*

Table 6.
Canonical correlation between data on meteorological variables and beetle population fluctuation record sets (U).

The Chi-Square (χ2) equal to 11.5400, with 8 degrees of freedom, was recorded in the Mata de Cocal, and the Chi-Square (χ2) equal to 10.6400, with 8 degrees of freedom, in the agricultural area, for the association between the variables sets: monthly records of the subfamilies population fluctuation and the monthly meteorological data. The Molytinae subfamily had a greater association with the monthly data set on meteorological variables in both areas.

The Curculionidae community in the agricultural area obtained the greatest associations with the monthly meteorological data as shown in **Table 6**. In both areas, there was association with the relative air humidity and precipitation positively and insolation and the average temperature in a negative way. In Mata de Cocal, there was a greater association with insolation (−0.80) and a lower association with temperature (−0.33), while there was a greater association with insolation and temperature (−0.87) and less association with precipitation (0.82) in the agricultural area.

Probably, the presence of trees in Mata de Cocal provided microtemporal conditions that favored the subfamilies not to become so dependent on meteorological variables.

3.3 Ecological niche of the family Curculionidae

The identified subfamilies specimens occupy a functional or biological position within the ecosystem in which they are inserted. This includes what they represent in the overall ecosystem, by what they do and how they do.

The curculionids, along with Cerambycidae family beetles, are the ones that occur most associated with native and exotic forest species, performing an important role in the wood degradation [28]. They stand out for the great number of species and high degree of polyphagia: there are species that can be xylophages, mycophages or spermatophytes [28–31]. They occur in forest species native to the Mata de Cocal, such as the babassu coconut palm [32]. These insects are common in tropical regions and only attack live trees that show changes in their physiological conditions [28, 31].

The curculionids diet basis is deficient in essential vitamins of group B and sterols, whose absence is compensated by a diet rich in nitrogen supplied by symbiotic fungi that synthesize them from nutrients absorbed from the galleries that they make inside the wood [30]. These specimens have a fundamental ecological role in the forests formation, as they recycle vegetable biomass. When they consume their hosts tissues, they facilitate the saprophytic organisms entry that accelerate the vegetable material deterioration [31].

Given that each species can be associated with several factors such as the climate, the soil, the vegetation type, among others, the deforestation processes of the native forest areas, can lead to the species loss, causing changes in their community [1].

All physical and biological entities in a given ecosystem form a unified and complex integral system [1]. As seen, there is a deep, direct and essential link between curculionids and the ecological processes of the ecosystem in which they are inserted.

Given this importance, it is necessary for them to remain in the environment in which they are, for that reason, there must be the maintenance of meteorological conditions related to insolation, precipitation, temperature and humidity, as well as the conservation – rational use – of the Mata de Cocal when its preservation is not possible – full and permanent protection.

4. Conclusion

Given the above, it is inferred that beetles populations certain (such as the subfamily Molytinae in the present study) are governed and conditioned by meteorological variables to a greater or lesser extent depending on the characteristics of the community itself and the biotic and abiotic environmental factors of the area where they live. From the case presented, it verifies that the Curculionidae community has a positive association with precipitation and humidity and a negative association with insolation and temperature, being that in native forests curculionids are not as dependent on meteorological variables as in agricultural fields.

Finally, as this chapter shows the association of curculionids with meteorological parameters in two habitats: Mata de Cocal and agricultural field; it becomes important as a parameter for a first description for further ecological studies of taxonomic refinements and deeper inflections.

Author details

Marcos Paulo Gomes Gonçalves
Federal University of Piaui (UFPI), Teresina, Brazil

*Address all correspondence to: mpgg1988@gmail.com

References

[1] Gomes Gonçalves M P. Relação Entre Tempo e Besouros em Mata de Cocal. Revista Brasileira de Meteorologia. 2017;32(4):543-554. DOI:10.1590/0102-7786324003

[2] Rodrigues W C. Fatores que influenciam no Desenvolvimento dos insetos. Info Insetos. 2004; 1(4):1-4.

[3] Jaworski T, Hilszczański J. The effect of temperature and humidity changes on insects development their impact on forest ecosystems in the expected climate change. Forest Rearch Papers. 2013;74(4):345-355. DOI:10.2478/frp-2013-0033

[4] França F M, Ferreira J, Vaz-de-Mello F Z, Maia L F, Berenguer E, Palmeira A F, Fadini R, Louzada J, Braga R, Oliveira V H, Barlow J. El Niño impacts on human-modified tropical forests: consequences for dung beetle diversity and associated ecological processes. Biotropica. 2020;52(2):252-262. DOI:10.1111/btp.12756

[5] Fernandes V, José P. A praga de gafanhotos no Sul da América: Argentina, Brasil e Uruguai (1890-1950). Journal of Social, Technological and Environmental Science. 2018;7(3):145-160. DOI:10.21664/2238-8869.2018v7i3.p145-160

[6] Lisbôa H, Grutzmacher A D, Wrege M S, Garcia F R M, Nava D E. Impact of global warming on the distribution of Anastrepha grandis (Diptera: Tephritidae) in Brazil. Arquivos do Instituto Biológico. 2020;87:1-10. DOI: 10.1590/1808-1657000882018

[7] Gallo D, Nakano O, Silveira-Neto S, Carvalho R P L, Batista G C, Berti-Filho E, Parra J R P, Zucchi R A, Alves S B, Vendramim J D, Marchini L C, Lopes S R S, Omoto C. Entomologia Agrícola. Piracicaba: FEALQ; 2002. 920 p.

[8] Ayres M P, Lombardero M J. Assessing the consequences of global change for forest disturbance from herbivores and pathogens. The science of the total environment. 2000;262(3):263-286. DOI: 10.1016/S0048-9697(00)00528-3

[9] Jönsson A M, Harding S, Bärring L, Ravn H P. Impact of climate change on the population dynamics of Ips typographus in southern Sweden. Agricultural and Forest Meteorology. 2007;146:70-81. DOI: 10.1016/j.agrformet.2007.05.006

[10] Netherer S, Schopf A. Potential effects of climate change on insect herbivores in European forests: general aspects and the pine processionary moth as specific example. Forest Ecology and Management. 2010;259:831-838. DOI: 10.1016/j.foreco.2009.07.034

[11] Gonçalves M P G. Os escaravelhos e o tempo. Revista Brasileira de Meteorologia. 2020. DOI:10.1590/0102-7786354034

[12] Gonçalves T T, Louzada J N C. Estratificação vertical de coleópteros carpófilos (Insecta: Coleoptera) em fragmentos florestais do sul do estado de Minas Gerais, Brasil. Ecol. Austral. 2005;15(1):101-110.

[13] Bernardes A C C, Oliveira O C C, Silva R A, Albuquerque P M C, Rebêlo J M M, Viana J H, Siqueira G M. Abundance and diversity of beetles (Insecta: Coleoptera) in land use and management systems. Revista Brasileira de Ciência do Solo. 2020;44:1-14. DOI: 10.36783/18069657rbcs20190183

[14] Gullan P J, Craston P S. Os Insetos: Um Resumo de Entomologia. Translated by Hoenen S. M. M. São Paulo: Roca; 2007. 440 p.

[15] Marinoni R C, Ganho N G. Fauna de Coleoptera no Parque Estadual de Vila Velha, Ponta Grossa,

Paraná, Brasil: abundância e riqueza das famílias capturadas através de armadilhas malaise. Rev. Bras. Zoologia. 2003;20(4):737-744. DOI: 10.1590/S0101-81752003000400029

[16] Santos S R de A. Diversidade de Coleoptera em um Fragmento de Mata Atlântica da Reserva Ecológica Gurjaú, Cabo de Santo Agostinho - PE [thesis]. Recife: Universidade Federal de Pernambuco; 2005.

[17] Rodrigues L R A. Os besouros coprófagos em pastagens. In: Anais do Simpósio sobre Ecossistema de Pastagens; Jaboticabal: FUNEP; 1989. p. 97-133.

[18] Paiva D R. Escarabeíneos (Coleoptera: Scarabaeidae) Associados a Diferentes Sistemas de Manejo de Pastagens no Município de Teresina-PI [thesis]. Teresina: Universidade Federal do Piauí; 2009.

[19] Campligia M. A Influência de Sistemas Silvipastoris sobre a Dinâmica Populacional de Besouros Coprófagos [thesis]. Florianópolis: Universidade Federal de Santa Catarina: 2002.

[20] Andrade Júnior A S, Bastos E A, Barros A H C, Silva C O, Gomes A A N. Classificação climática e regionalização do semiárido do Estado do Piauí sob cenários pluviométricos distintos. Revista Ciência Agronômica. 2005;36(2):143-151.

[21] INSTITUTO BRASILEIRO DE GEOGRAFIA E ESTATÍSTICA (IBGE). Manual Técnico da Vegetação Brasileira. Série Manuais Técnicos em Geociências. n. 1. Rio de Janeiro: Ed. IBGE; 2012. 271 p.

[22] Santos-Filho F S, Almeida Jr. E B, Soares C J R S, Zickel C S. Fisionomias das restingas do Delta do Parnaíba, Nordeste, Brasil. Revta. Bras. Geog. Física. 2010; 3(3):218-227. DOI: 10.26848/rbgf.v3.3.p218-227

[23] Farias R R S, Castro A A J F. Fitossociologia de trechos da vegetação do Complexo Campo Maior, PI, Brasil. Acta. Bot. Bras. 2004;18(4):949-963. DOI: 10.1590/S0102-33062004000400025

[24] Costa E M, Araújo E L, Lopes da Silva F E, Nogueira C H F, Silva P A F Diversidade de coleópteros em área cultivada com melancia no semiárido do rio grande do norte. Revista Agro@mbiente On-line. 2014;8(2):293-297. DOI: 10.18227/1982-8470ragro.v8i2.2058

[25] Moraes R C B, Haddad M L, Silveira Neto S, Reyes A E L. Software para análise faunística. In: Anais 8° Simpósio de controle biológico; S. Pedro: 2003. p. 195.

[26] Ayres M, Ayres-Júnior M, Ayres D L, Dos Santos A S. BioEstat: Aplicações Estatísticas nas Áreas das Ciências Biológicas e Médicas. Belém: Publicações Avulsas Mamirauá. 2007. 364 p.

[27] Bastos E A, Andrade-Júnior A S. Boletim agrometeorológico de 2010 para o município de Teresina, PI. Embrapa Meio Norte série documentos-210. 2011.

[28] Gray B. Economic tropical forest entomology. Annual Review of Entomology. 2003;17(1):313-352. DOI: 10.1146/annurev.en.17.010172.001525

[29] Castrejón A J E, Montesinos M R, Acevedo R N, Tamez G P, Ayala Z M Á, Berlanga P A M. Especies de Xyleborus (Coleoptera: Curculionidae: Scolytinae) asociados a huertos de aguacate en Colima, México. Acta Zool. Mex. 2017;33(1):146-150.

[30] Ranger C M, Reding M E, Schultz P B, Oliver J B, Frank S D, Adesso K M, Chong J H, Sampson B, Werle C W, Gill S, Krause C. Biology, Ecology, and Management of Nonnative Ambrosia Beetles (Coleoptera: Curculionidae: Scolytinae) in Ornamental Plant Nurseries. Journal of Integrated Pest Management. 2016;7(1);1-23. DOI: 10.1093/jipm/pmw005

[31] Smith S M, Petrov A V, Cognato A I. Beetles (Coleoptera) of Peru: a survey of the families. Curculionidae: Scolytinae. The Coleopterists Bulletin. 2017;71(1):77-94. DOI: 10.1649/0010-065X-71.1.77

[32] Takada H M, Batista-Filho A, Hojo H. Aspectos biológicos de Rhynchophorus palmarum Linnaeus, 1764 (Coleoptera: Curculionidae) em banana-prata (*Musa sapientum*). Biológico. 2014;76(1):35-39.